Wellington Gomes da Silva
Fátima C. Rezende
Jacinto A. Carvalho

Irrigation management for meter beans

Wellington Gomes da Silva
Fátima C. Rezende
Jacinto A. Carvalho

Irrigation management for meter beans

Technical and economic analysis

ScienciaScripts

Imprint

Cover image: www.ingimage.com

This book is a translation from the original published under ISBN 978-613-9-67018-5.

Publisher:
Sciencia Scripts
is a trademark of
Dodo Books Indian Ocean Ltd. and OmniScriptum S.R.L publishing group

120 High Road, East Finchley, London, N2 9ED, United Kingdom
Str. Armeneasca 28/1, office 1, Chisinau MD-2012, Republic of Moldova, Europe
Printed at: see last page
ISBN: 978-620-8-10585-3

SUMMARY

To my parents, Mr. Antônio and Mrs. Antônia, for their examples of humility.

To my siblings Eleud, Jaqueline, Katiane, Poliana and Samirimis, and other family members, for their declarations and/or demonstrations of love, encouragement and unconditional support.

Especially my son Wellington, the main reason for my incessant search for better conditions and quality of life.

DEDICATE

ACKNOWLEDGMENTS

To God, for the divine grace of my existence, for my health and for allowing me to realize this dream of mine, which at times seemed distant and unattainable.

To the Federal University of Lavras (UFLA) and the Department of Engineering (DEG), in the person of Professor Dr. Manoel Alves de Faria (then coordinator of the Postgraduate Program in Agricultural Engineering), and other members of the collegiate, for the opportunity offered to carry out the Doctorate course.

To the Federal University of Amazonas (UFAM), for the opportunity and support to train in a Postgraduate Program at an Institution outside its headquarters.

To the Research Support Foundation of the State of Minas Gerais (FAPEMIG), for the financial support, which was essential for carrying out the experiments.

To the Amazonas State Research Support Foundation (FAPEAM), for granting the scholarship, which was essential to cover the cost of teaching materials, travel, housing, food and educational expenses.

To the professors at UFAM, Dr. Eduardo Ossamu Nagao, Dr. Carlos Alberto Franco Tucci, Dr. Carlos Moisés Medeiros, Dr. Antenor Francisco de Figueiredo and Dr. Edinaldo Narciso Lima, for their friendship, support and encouragement to complete the PhD course.

To Professor Dr. Antônio Marciano da Silva, for his support and encouragement during my doctoral studies.

To Professor Dr. Jacinto de Assunçâo Carvalho, for his guidance, support in conducting the experiments, for his understanding, patience, dedication and promptness.

To the researcher Dr. Fâtima Conceição Rezende, for her prompt attention, determination in her intense and dedicated co-supervision, evaluation and suggestions that contributed to improving the quality of the thesis.

To Professor Dr. Geraldo Magela Pereira, for his support in co-supervision, prompt attention, evaluation and suggestions that contributed to improving the quality of the thesis.

To Professor Dr. Messias José Bastos de Andrade for his contribution in co-supervising the experiments.

To the researcher Dr. Myriane Stella Scalco, for her promptness, collaboration in the evaluation and suggestions that contributed to improving the quality of the thesis.

To Professor Dr. Joelma Pereira Durâo, for her collaboration in the evaluation and suggestions that contributed to improving the quality of the thesis.

To all the professors in the Engineering Department with whom I have had the opportunity to study subjects, for their dedication, seriousness and commitment in teaching.

To the technical-administrative staff and friends of the Water and Soil Engineering Sector, Mr. José Luiz and Osvaldo "Neném", for having contributed directly and/or indirectly to my professional training and for their valuable help in my research activities.

To Dr. Nerci Nina Lima and other family members, for the warmth with which they welcomed me as a guest in their home during the initial period of my arrival in Lavras.

To the Agricultural Engineer MSc. Carla de Pàdua Martins, for her affection, friendship, companionship, hospitality and support at all times.

To the Agricultural Engineer MSc. Arionaldo de Sâ Jûnior, for the pleasant daily interaction, friendship and support during the period when I fractured my arm and was hospitalized.

To the Agricultural Engineer MSc. Eduardo Carvalho Oliveira, for the pleasant coexistence, friendship, help in understanding and solving the academic exercises of the subjects we studied together.

To the Agricultural Engineer MSc. Gervâsio Rios, for the pleasant coexistence, friendship, help in understanding academic theories and concepts and solving the exercises of the subjects we studied together, support in conducting the experiments and during the period when I was hospitalized.

To the Agricultural Engineer MSc. Donizete dos Reis Pereira, for the pleasant interaction and friendship.

To Bruno Montoani Silva, an agricultural engineer, for his friendship and help in conducting the experiments.

My most sincere thanks go to all those who contributed in some way to the completion and improvement of this work.

"The individual is, as the word indicates, indivisible. He carries wherever he goes what he is."

John Dewei

SUMMARY

The production of subway beans (*Vigna unguiculata* spp. *sesquipedalis* (L.) Verdc.) is an excellent alternative source of income for producers in the state of Amazonas, since they are widely consumed in the region. However, there is practically no information in the literature on the behavior of the crop when grown under different soil moisture conditions and irrigation regimes in a protected environment. The general aim of this work was to obtain parameters for irrigation management in the vegetative and productive phases of the subway bean and to assess the economic viability of growing it in a protected environment. Two experiments were carried out with the crop in a protected environment. The first experiment was a randomized block design with a 5x2 factorial scheme, corresponding to five levels of matric potential (equivalent to -15, -35, -55, -75 and -95 kPa), and two phenological phases of the crop (vegetative and reproductive). The second experiment was also carried out in a randomized block design, but with varying levels of water replenishment (40, 70, 100, 130 and 160%). Four replications per treatment were used in each experiment. Development (height, diameter and dry matter), yield (number of pods, production and productivity) and fruit quality (fresh mass and pod length) were assessed. The data was submitted to analysis of variance and regression. It was concluded that it is economically feasible to grow metro beans in a protected environment and that irrigation management should be carried out while maintaining soil moisture close to field capacity (- 15 kPa) in both phenological phases of the crop.

Keywords: *Vigna unguiculata* spp. *sesquipedalis.* Irrigation. Economic analysis. Phenological phase.

PART ONE

1 INTRODUCTION

The meter bean, from the Fabaceae family (*Vigna unguiculata* spp. *sesquipedalis* (L.) Verde.), is much appreciated by the populations of the North and Northeast of Brazil and is grown by small producers. Due to its hardiness and tolerance to high temperatures, it has become an option for pod bean cultivation (SILVA, 2003). It has a high protein content (28%), iron (2.5 mg 100 g^{-1}), calcium (80 mg 100 g^{-1}), phosphorus (74 mg 100 g^{-1}), vitamin A (941 IU 100 g^{-1}), vitamin C (13mg 100g^{-1}) and fiber (2g 100g^{-1}), according to RESMI and GOPALAKRISHNANO (2004).

This species represents a significant alternative source of income for rural producers in Amazonas, but it is most intensively cultivated in the initial period of the rivers' ebb and flow, when farmers take advantage of the natural humidity of the lowland soils. Thus, its supply is seasonal and dependent on favorable natural conditions (CARDOSO, 1997).

The instability of the supply of agricultural products, resulting from variations in climatic elements, is contributing to an increase in the use of cultivation techniques that allow for greater stability in production; among these are irrigation and cultivation in a protected environment.

Irrigation is commonly used by farmers to maintain a regular supply of water for cultivated species. Combined with other agricultural practices, irrigation becomes a factor in guaranteeing both a stable supply of products and an increase in crop productivity.

However, this practice will only bring benefits if it is done rationally, minimizing expenditure on water, energy, equipment, fertilizers and other elements that are factors in production costs (FRIZZONE, 2007a).

By knowing, among other factors, the water requirement and the ideal time to supply water to the species of interest in its different stages of development, rational irrigation management can be established. Only in this way can economically viable and profit-maximizing production be achieved.

Cultivation in a protected environment is not new. It was introduced to Brazil in the 1950s by the Dutch for flower production, and in the 1970s it began to be used for growing vegetables (MARTINS, 1996a). However, there is no information on management under these growing conditions for all crops of agricultural interest.

According to SGANZERLA (1995), growing crops in a protected environment protects the plants from weather agents, shortens the production cycle, increases productivity, improves product quality and allows harvests to be obtained at any time of the year, thus guaranteeing stability in the supply of produce from the species grown in this system.

There is still no research that provides subsidies for technically and/or economically viable irrigation management for subway beans in this production system.

In view of the above, the general aim of this work was to obtain parameters for irrigation management in the vegetative and productive phases of subway beans and to assess the economic viability of growing them in a protected environment.

2 THEORETICAL FRAMEWORK

2.1 Growing subway beans

The subway bean (*Vigna unguiculata* spp. *sesquipedalis* (L.) Verde.) is a vegetable widely grown in the world. It originated in Central Africa, but was first cultivated in Southeast Asia (HA et al., 2010). In Brazil, this crop was introduced to the Northeast by European colonizers, possibly in the state of Bahia. Subsequently, it spread to other regions of the country, mainly to the North and Midwest. In the North, it is widely consumed, mainly because it replaces the "pod" in various dishes, especially in salads (CARDOSO, 1997).

The plant is a liana (climber), reaching approximately 2.0 m in length. The root is pivotal and, like the stems, has few branches. The leaves have a very peculiar morphology, typical of species belonging to the Fabaceae family. They are compound and trifoliate, long-petiolate, alternate and stipulate. The inflorescence is cluster-shaped, with violaceous, long-stalked flowers (PIMENTEL, 1985). Due to the beauty of its flowers, this species is also used as an ornamental plant.

According to OFORI and KLOGO (2005), metro bean pods should be harvested at an immature stage, before the seeds have fully developed. According to RESMI and GOPALAKRISHNAN (2004) the start of harvest can be estimated by counting the number of weeks after planting. According to these authors, it takes around seven weeks from sowing to the start of harvest, depending on the cultivar and environmental conditions; the harvest period normally lasts six to eight weeks. According to USAID (2004), the length and diameter of the pods and the number of days after anthesis (flower opening) are the main indicators of when to harvest.

Although it is recommended to harvest the pods at a minimum length of 38 cm, some markets prefer longer pods, up to 76 cm. The best quality pods are straight, uniform in length and homogeneous in color. The most popular cultivars are green in color, although some markets may prefer cultivars that produce a reddish pod (USAID, 2004).

In a study with irrigated subway beans under field conditions carried out at the University of Ghana (West Africa), OFORI and KLOGO (2005) concluded that, in order to obtain pods of good quality for consumption, they should be harvested before 15 days after the flower opens.

Based on the quotes above, there are two criteria that can be adopted for harvesting. The first, based on the length of the pods, seems to be more practical for producers to adopt. The second criterion, which recommends harvesting based on the number of days after anthesis, may be more interesting when conducting research, as it may better represent the plants' response to variations in the environmental conditions to which they are subjected. The adoption of one or other harvesting method will depend on the researcher's objectives.

In Asian countries such as China, India, Malaysia and Indonesia, for example, where the subway bean is of greatest economic interest, there are more than 30 commercial varieties. In the northern region of Brazil, only two varieties are grown: one with thin pods and black seeds and the other with thicker pods and brown seeds. In this region, consumer preference is indistinct between the two types (PIMENTEL, 1985).

Because the plant is a climber, it needs to be supported, which can be done using espaliers, small diameter poles or even bamboo. The choice of the type of support will depend on the availability of material and the growing environment (SGARZERLA, 1997).

Demand in the northern region is mainly met by small producers who have this vegetable as a significant source of income. In the state of Amazonas, where consumption is widespread, plantations by farmers in the vàrzea area predominate (CARDOSO, 1997).

Despite its economic and social importance, subway beans are still produced seasonally, as their cultivation depends on favorable natural conditions; as a result, their supply does not meet the demand of the regional market all year round (PIMENTEL, 1985).

In field conditions, subway beans are planted at the end of the rainy season, when the plant has water at its disposal at the start of growth and fruiting is not affected by high rainfall, which compromises production (PIMENTEL, 1985). Thus, the cultivation of this vegetable is restricted to this period as it is the time that provides the most favorable environmental conditions for production.

According to PIMENTEL (1985), it is recommended to use irrigation to grow subway beans during the dry season. In a protected environment, the only source of water for the crop comes from irrigation (PARIZI et al., 2010). However, research results on irrigation management for the crop are scarce.

Irrigation management and cultivation in a protected environment, in addition to increasing productivity, can provide a differentiated product, of better quality and with the prospect of better prices on the market (MARTINS et al., 1995b). As this is a practice that has not yet been adopted in the cultivation of meter beans, the promising prospect of adopting irrigation should be studied and analyzed in detail, especially with regard to management and economic viability.

2.2 Cultivation in a protected environment

The main difference between agricultural production and other sectors of the economy is the exposure to environmental conditions. Agriculture is a high-risk activity, since there is no control over climatic elements (ROMANINI et al., 2010). Modern agriculture produces sustainably with the lowest possible degree of risk, through the intensive use of appropriate techniques, such as protected cultivation and irrigation (ARAÙJO et al., 2009). According to MARY et al. (2007), growing crops in a protected environment is a technique that has contributed to the modernization of agriculture.

The use of a protected environment allows crops to grow and develop at times when the meteorological elements in the external environment are limiting (HELDWEN et al., 2010), 2010), greater protection from climatic factors (excessive rainfall, lower night-time temperatures, etc.), protection of the soil from leaching, reduced costs of fertilizers and pesticides, and higher yields compared to open-air cultivation (REIS et al., 2009), production at times of better prices and better product quality.

What is expected from a crop grown in a given environment is high yield and product quality, provided that the crop is given the conditions to use the factors that determine production as efficiently as possible, whether they are genetic, biotic or abiotic (SGANZERLA, 1997).

Conditions of solar radiation, temperature and relative humidity can be modified by the use of a protected environment (SGANZERLA, 1997).

According to NASCIMENTO et al. (2008), an increase in air temperature affects flowering potential, reducing the number of days until flowering begins. According to these authors, this is due to its effect on the plant's internal metabolic processes.

CARVALHO et al. (2010a) found that there was no effect of the irrigation regimes in the protected and natural environments up to a matric potential of -60 kPa, but they did observe that the passion fruit plants (*Passiflora edulis*) grew faster in height and diameter, had an earlier harvest and also had better quality pods compared to the natural environment. According to the authors, the faster growth of the plants may have been influenced by the higher maximum and minimum temperatures in the protected environment compared to the natural environment.

Reductions of around 60% in crop evapotranspiration have been observed in plastic-covered houses compared to open-air crops (MARTINS and GONZALEZ, 1995c). According to FARIAS et al. (1994), plastic cover reduces atmospheric evaporative demand, especially due to the attenuation of radiation and the reduction in wind speed, which are considered to be the most important factors in atmospheric evaporative demand.

2.3 Irrigation management

The demand for water resources in agriculture, due to the incorporation of new areas into the production process, both in Brazil and elsewhere in the world, has increased concerns about the rational use of water (COELHO and OR, 1999).

As far as irrigated agriculture is concerned, water is not used as efficiently as expected. Low efficiency is observed not only in its conduction and distribution by systems, but also in irrigation management (CHRISTOFIDIS, 2006). There is therefore a need for greater efforts to make more efficient use of this natural resource.

Existing irrigation systems include conventional sprinkler irrigation, self-propelled irrigation, center pivot irrigation and drip irrigation. It is important to note that there is no one method that is better than the other in terms of plant physiology, but they differ in their adaptability in relation to local soil conditions, topography, climate, crop, water quality, economic factors and certain external and agronomic influences (OLIVEIRA et al., 2010a). Each system has its advantages and disadvantages, but the drip system is the most suitable for protected environments.

Regardless of the irrigation system used, it is essential to manage irrigation properly in order to rationalize the use of all the factors involved in the production process with a view to obtaining a better return on the capital employed (FRIZZONE and ANDRADE

JÛNIOR, 2005b).

According to SOUZA et al. (2006a), in order for irrigation management to be carried out satisfactorily, it is necessary not only to have the means to help determine when to irrigate and how much water to apply, but also to know how crops respond to these procedures.

Normally, excess or deficit water, two conditions that characterize the low efficiency of water use in irrigated farming systems, are limiting factors in obtaining high yields and good quality products (PIRES et al., 2007b).

Poor irrigation management results in losses related to excessive spending on soil fertilization, due to nutrient leaching and their low availability to the plant; energy losses from pumping water unnecessarily; soil salinization and other resulting complications (COSTA et al., 2002; FARIA et al., 2002).

Although irrigation management is a factor of great importance for those who wish to succeed in farming, it must be associated with other agricultural practices (soil preparation, fertilization and liming, use of improved seeds, spacing, planting density, pest and disease control) so that the efficiency of water use is maximized (SOUZA et al., 2005b), as it does not produce results on its own.

Irrigation management can be carried out based on monitoring soil water potential and reference evapotranspiration obtained by the Penman-Monteith method, modified by the FAO or by the "Class A" tank, among others. According to BLAINSKI et al. (2009), monitoring based on soil water potential is the most widespread and simplest to operate, although it has disadvantages just like the other methods.

LOPES et al. (2004), evaluating the effect of irrigation management by monitoring soil water potential and by the "Class A" tank method on plant water consumption and grain yield of the bean crop, concluded that there were water savings of around 15% when management based on soil water potential was used, without affecting grain yield. These results reinforce the claims made by other authors (PHENE and HOWELL, 1984; OLIVEIRA et al., 1999a) that water management based on soil moisture is more efficient than other methods.

The measurement of the matrix potential has been the subject of research in the search to establish the management of irrigation water in lettuce crops (ANDRADE JÛNIOR and KLAR, 2003; SANTOS and PEREIRA, 2004; MAGGI et al., 2006), carrots (SILVA et al., 1982b), peppers (SANTANA et al...), beans (ESPINOSA, 1980; FIGUEIREDO, 1998a; GUERRA et al., 2000a), 2004), beans (HOSTALACIO E VALIO, 1984; FIGUEIREDO, 1998a; GUERRA et al., 2000a), wheat (ESPINOSA, 1980; GUERRA, 1995b); sorghum (COGO E GUERRA, 1978), cotton (OLIVEIRA et al., 1999b), melons (BRAGA et al., 2006), strawberries (PIRES et al., 2007b), passion fruit (CARVALHO et al., 2010a), among others.

The total amount of water required to meet the crop's needs during its cycle must be distributed in such a way as to meet the differential water demand of the plants at each phenological stage. This increases from seed soaking, which requires very low values, to the point of maximum leaf area where there is maximum evapotranspiration (LARCHER, 2000).

This difference in the water demand of crops is related to the plant's critical period for water deficit. According to LARCHER (2000), the critical period for a certain factor for the plant (temperature, rainfall, humidity, etc.) is an interval during which it is most susceptible. According to SILVA et al. (2009), during critical periods for soil moisture, water deficiencies reduce economic yields more than in other periods.

Thus, by defining the critical period for crops to suffer from water deficit, in addition to improving the efficiency of the water used in irrigation, production costs can be reduced due to lower expenditure on water, energy and labor used to maintain the irrigation system (FRIZZONE, 2007a).

As a result, several studies have been carried out in order to define the irrigation water management for the respective phenological phases of the crops, among which we highlight the research carried out by STONE et al., 1979b; ABREU et al., 1980; GUERRA et al., 1995b; MAROUELLI et al., 1991a, 2002b 2003c; SA et al., 2005 and BILIBIO et al., 2010a.

2.4 Matrix potential

Soil water can be characterized by a state of energy, which is determined according to the different forces acting on it (LIBARDI, 2005). According to HILLEL (2004), classical physics recognizes the two main ones as kinetic and potential. According to this author, as the movement of water in the soil is very slow, the contribution of kinetic energy is insignificant. On the other hand, as water moves following an energy gradient created by the action of certain forces related to the soil, potential energy becomes the main one. This state of energy is therefore called the potential of water in the soil (REICHARDT and TIMM, 2004).

JURY and HORTON (2004) state that the water potential in the soil is often determined by the sum of its partial components (gravitational potential, matric potential, pressure potential and osmotic potential). However, for PIRES (2002a) the matric potential has been, in practical terms, the only one to be considered in irrigation management, when this is done on the basis of soil moisture control.

PREVEDELLO (1996) reports that the lower the matric potential (more negative) adopted to define the moment of irrigation, the less frequent the repositioning of water in the soil for the crop, and vice versa. According to KLAR (1984), this is because irrigations are carried out when the soil is most depleted of available water, which makes the period between irrigations longer.

STONE et al. (1988a), when studying the effects of three levels of matric potential (-12.5 kPa, -25 kPa and -75 kPa) on bean growth, found that there was a considerable decrease in growth as matric potential decreased. As irrigation was carried out at lower matric potential, leaf area index and duration, yield and dry matter production rate, relative growth rate and net assimilation rate decreased. According to these authors, the reduction in dry matter production was almost exclusively due to the leaf area index since water stress, by reducing turgidity, reduces cell expansion, which in turn reduces stem and leaf expansion.

Research has shown that, in general, crop yields increase when irrigations are carried out at higher values of matric potential (close to field capacity), i.e. more frequent irrigations (ANDRADE JÛNIOR E KLAR, 2003a; SANTOS e PEREIRA, 2004; FIGUERÊDO, 2008a).

In irrigation management studies based on matric potential, SANTANA et al. (2004) found in an experiment with pepper crops, with irrigation restarted at -10, -30, -50 and -60 kPa, that the best yield averages were achieved with the treatment whose water was replaced when the soil humidity reached -10 kPa, which indicates a higher matric potential.

On the other hand, SA et al. (2005) concluded that the highest tomato yields were obtained by irrigating the plants when the matrix potential reached values close to -80 kPa, while irrigations at matrix potentials of -15, -30, -50, -120 and -170 kPa reduced yields.

As you can see, the highest yields are not always obtained with more frequent irrigations. These results clearly show that irrigation management cannot be generalized.

2.5 Economic analysis

In view of the drop in proclucTion of the meter bean under field conditions, caused by high rainfall (PIMENTEL, 1985), the interruption of its supply to the consumer market during the least favorable periods for planting and its social and economic importance, it is possible that growing this vegetable, irrigated and in a protected environment, can provide income generation for producers throughout the year, as well as promoting stability in the supply of produce to consumers. However, economic aspects must be taken into account when recommending this cultivation practice, as one cannot suggest spreading agricultural practices based solely on agronomic results.

The importance of economic analysis studies has increased in recent years due to the tendency for the price of inputs to increase in relation to products. Hence the need for farmers to use these inputs efficiently in order to minimize variable costs per unit produced (FRIZZONE and ANDRADE JÙNIOR, 2005b).

The share of irrigation in production costs is an important factor. Furthermore, economic analysis is fundamental in the decision-making process of whether or not to grow meter beans in a protected environment (FRIZZONE and ANDRADE JÙNIOR, 2005b).

The economic evaluation of irrigation often involves quantifying productivity in response to the total amount of water applied. The characterization of the crop's response to water application has been widely known as the Water-Crop Production Function (SOUZA et al., 1996c). The use of production functions in these studies is justified by the need to find an agroeconomic indicator without the need to resort to complex methods that are difficult to operate (FRIZZONE and ANDRADE JÙNIOR, 2005).

These functions are empirical relationships obtained through regression analysis between productivity (dependent variable) and the amount of water applied (independent variable), according to a given model (FERGUSSON, 1998).

The model representing the plant response function to irrigation water management should be chosen on the basis of its ability to best represent the experimental data, the ease with which the model can be adjusted to the data and the interpretation of the results (1984).

The model that describes the water-crop production functions, obtained experimentally, can indicate constant returns with the increase in water levels applied. This means that there will be constant productivity when the water levels applied result in equal increases in the productivity obtained. Thus, the relationship between water application and productivity will be linear. When this happens, it is not possible to determine the optimum economic combination between the variable factors and the fixed factors used in the production process (ALMEIDA et al., 2004).

AZEVEDO et al. (2005), studying the yield of pepper trees (*Capsicum* spp.) as a function of the management carried out with the application of increasing water levels (674, 765, 873, 969 and 1,080 mm), found that the mathematical representation of the production function followed a linear model, which means constant increases in productivity as a function of the water levels applied. These results make it impossible to carry out an economic analysis according to the principles of the economic theory of production, as FERGUSON (1998) points out.

Among the mathematical models used in the economic analysis of agricultural research, the second-order polynomial stands out, since most of the time it is the one that has best represented the production estimate in production function studies in irrigated agriculture (FRIZZONE et al., 1987c; ANDRADE JÙNIOR, 2002; ALMEIDA, 2004; MONTEIRO et al., 2006; VILAS BOAS et al., 2008; SANTANA et al., 2009; BILIBIO et al., 2010b; CARVALHO et al., 2011b, OLIVEIRA et al., 2011).

The national scientific literature contains economic analysis studies for a large number of crops of agricultural interest which identify, using crop response functions to water application, the level of water to be used in irrigation management and the yield that can provide the best economic return when using irrigation for the crop of interest (SANTANA et al., 2009; CARVALHO et al., 2011b, OLIVEIRA et al., 2011).

However, the adoption of the recommended irrigation management based solely on the indication of the optimum economic blade, estimated from the production functions obtained experimentally, does not give the producer any idea of the financial return to be obtained from the decision to invest in this agricultural activity. According to QUEIROZ et al. (1996) the use of irrigation can become economically unviable, even with the application of the optimum economic rate, if production costs are not taken into account during the economic analysis.

There is therefore a need to include some economic viability indicators in these studies, based on which it is possible to predict the level of financial return on the capital to be employed, thus contributing to a better analysis for decision making by rural producers.

Various researchers (MARQUES and COELHO, 2003; SOUZA et al., 2009b; PONCIANO et al., 2006; PIMENTEL et al., 2007; CECiLIO FILHO et al., 2008; PAVINATO et al., 2008; AREDES et al., 2009) used economic viability indicators to gauge the likelihood of success or failure in adopting production systems in agricultural activities, including the Net Present Value (NPV), the Internal Rate of Return (IRR) and the Benefit/Cost ratio (B/C).

For the cultivation of irrigated meter beans in a protected environment to be economically viable, the economic benefits generated by the adoption of this cultivation technique must be positive and higher than the costs, i.e. the monetary benefit generated by production must be greater than the increase in production costs caused by the implementation and operation of the irrigation system (FRIZZONE and ANDRADE JÛNIOR, 2005b).

REFERENCES

ARAUJO, Jucilene S. et al . Fruit characteristics of peppers grown in a protected environment under doses of nitrogen via fertigation. **Revista brasileira de engenharia agricola e ambiental**, Campina Grande, v. 13, n. 2, 2009 .

ABREU, T. A. S. et al. Analysis of onion production under different irrigation regimes. **Pesquisa agropecuâria brasileira**, Brasilia, v.15, n.2, p.233-236. 1980.

ALMEIDA, Economic analysis based on response functions of productivity versus water levels for papaya in northern Rio de Janeiro. **Engenharia Agricola**, Jaboticabal, v. 24, n. 3, p. 675-683, 2004.

ANDRADE JÙNIOR, A. S.; KLAR, A. E. Lettuce productivity as a function of soil water potential and irrigation levels. **Horticultura Brasileira**, v. 14, n. 1, p. 27-31, 2003a.

ANDRADE JÙNIOR, et al. Irrigation levels in cowpea cultivation. **Revista Brasileira de Engenharia Agricola e Ambiental**, v.6, n.1, p.17-20, 2002b.

AREDES, A. F. et al. Economic analysis of irrigation in passion fruit cultivation.

Revista de Economia da UEG, Anâpolis, v. 5, n.1, 2009.

AZEVEDO, B. M. de et al. Yield of pepper trees as a function of irrigation rates. **Ciência Agronômica**, v. 36, n. 3, p. 268-273, 2005.

BILIBIO, Carolina et al . Vegetative and productive development of eggplant subjected to different soil water tensions. **Revista brasileira de engenharia agricola e**

ambiental, Campina Grande, v.14, n. 7. 2010a.

BILIBIO, C.; CARVALHO, J. A.; MARTINS, M.; RESENDE, F. C.;

FREITAS, W. A.; GOMES, L. A. A. Production function of eggplant irrigated in a protected environment. **Irriga**, Botucatu, v. 15, n. 1, p. 10-22, 2010b. BLAINSKI, E. et al . Optimum hydric interval in an irrigated red nitossolo distroférrico. **Revista Brasileira de Ciência do Solo**, Viçosa, v. 33, n. 2. 2009.

BRAGA, M. B. et al. Geographical orientation of polyethylene greenhouses and soil water potentials in the cultivation of lacy melon hybrid 'nero'. **Irriga, Botucatu,** v. 11, n. 1, p. 130-138. 2006.

CARDOSO, M.O. **Unconventional vegetables from the Amazon**. EMB RAPA- MA. Brasilia - DF, 1997.

CARVALHO, J. A. et al . Development and productivity of yellow passion fruit irrigated under different irrigation rates in protected and natural environments. **Engenharia Agricola**, Jaboticabal, v.30, n.5. 2010a.

CARVALHO, J. A. et al. Productive and economic analysis of red peppers irrigated with different rates, grown in a protected environment. **Revista Brasileira de engenharia agricola e ambiental**, Campina Grande, v. 15, n. 6, p. 569-574, 2011b.

CECILIO FILHO, A. B. et al . Productive and economic viability of the consortium between chicory and arugula as a function of planting time. **Horticultura Brasileira**, Brasilia, v. 26, n. 3, 2008.

COELHO, E. F.; OR, D. Model of water distribution and matric potential in soil under drip irrigation with water extraction by roots. **Pesquisa agropecuària brasileira,** Brasilia, v.34, n.2, p. 227-234. 1999.

COSTA, E. L. et al. Irrigation of onions. **Informe agropecuàrio**, Belo Horizonte, v. 23, n.218, p. 57-66. 2002.

CHRISTOFIDIS, D. Water in food production: the role of academia and industry in achieving sustainable development. **Revista ciências exatas**, Taubaté, v. 12, n. 1, p.

37- 46, 2006.

ESPINOSA, W.; SILVA, E. M; SOUZA, O. C de. Model of water distribution and matric potential in soil under drip irrigation with root water extraction. **Pesquisa agropecuària brasileira.** Brasilia, v.15, n.1, p. 107-115, 1980.

FARIA, R. A. et al. Water and energy savings in supplementary irrigation projects in the state of Minas Gerais. **Revista Brasileira de Engenharia Agricola e Ambiental**, Campina Grande, v. 6, n. 2, p. 189-194, 2002.

FARIAS, J. R. B. et al. Effects of plastic cover on solar radiation.

Revista Brasileira de Agrometeorologia, v.1, n.1, p.31-36, 1994.

FERGUSON, C. E. **Microeconomia**. 13 ed. Rio de Janeiro: Forense Universitaria, 1998. 610p.

FIGUERÊDO, S. F. et al. Bean irrigation management based on technical and economic criteria in the cerrado. **Irriga**, Botucatu, v. 13, p. 378-391, 2008a.

FIGUERÊDO, S. F. et al. Establishment of irrigation timing based on soil water tension for the bean crop. **Engenharia Rural**, Piracicaba, v. 9, p. 35-49, 1998.

FRIZZONE, J. A. Irrigation planning using optimization techniques. **Revista Brasileira de Agricultura Irrigada**, Fortaleza, v.1, n.1, p.24-49. 2007.

FRIZZONE, J. A.; ANDRADE JUNIOR, A. S. de (Eds.). **Irrigation planning**: investment decision analysis. Brasilia, DF: EMBRAPA Informaçâo Tecnológica, 2005. 626p.

FRIZZONE, J. A.; OLLITA, A. F. L.; PEREIRA, G. T. Bean response functions to the use of nitrogen and irrigation: 1- rational region.

Irrigation and Modern Technology, v. 28, 1987.

GUERRA, A. F.; SILVA, D. B. da; RODRIGUES, G. C. Manejo de irrigaçâo e fertilizaçâo nitrogenada para o feijoeiro na regiâo dos cerrados. **Pesquisa agropecuâria brasileira**, Brasilia, v. 35, n. 6. 2000.

GUERRA, A. F. Soil water tension: effect on barley grain yield and quality. **Pesquisa**

agropecuâria brasileira, Brasilia, v.30, n.2, p.245-254. 1995.

HA, T. J. et al. Identification and Characterization of Anthocyanins in Yard- Long Beans (Vigna unguiculata ssp. sesquipedalis L.) by High-Performance Liquid Chromatography with Diode Array Detection and Electrospray Ionization/Mass Spectrometry (HPLC-DAD-ESI/MS) Analysis. **Journal of Agricultural and Food Chemistry**, [S. l], v. 58, n. 4, p. 2571-2576. 2010. 2010.

HELDWEIN, A. B. et al . Plastochron and yield of cowpea grown under protected environment and outdoors in late fall sowing. **Ciencia Rural**, Santa Maria, v. 40, n. 4. 2010.

HILLEL, D. **Introduction to environmental soil physics.** Amsterdam: Boston, MA: Elsevier Academic Press, 2004. 494 p.

HOSTALCIO, S.; VALIO, I. F. M. Development of early Goiano bean plants under different irrigation regimes. **Pesquisa agropecuâria brasileira**, Brasilia, v.19, n.2, p.211-218. 1984.

JURY, W.; HORTON, R. **Soil physics**. 6th. ed. New Jersey. Hoboken: J. Wiley, 2004. 370 p.

KLAR, A. E. **Water in the soil-plant-atmosphere system.** Sâo Paulo: Nobel, 1984 408 p.

LARCHER, W. **Plant ecophysiology**. Sâo Carlos: Rima Artes e Textos, 2000. 531p.: ill.

LIBARDI, P.L. **Dynamics of water in the soil**. Sâo Paulo, Editora da Universidade de Sâo Paulo, 2005. 327p.

LOPES, A. S. et al. Irrigation management (tensiometry and climatological water balance) for the bean crop in direct and conventional cultivation systems. **Engenharia Agricola**, Jaboticabal, v. 24, n.1, p.89-100, 2004.

MAGGI, M. F. et al. Production of lettuce varieties under different soil water potentials in a protected environment. Irriga, Botucatu, v. 11, n. 3, p. 415-427, 2006.

MARY, W. et al. Evaluation of a bamboo structure as a construction element for a greenhouse. **Engenharia Agricola**, Jaboticabal, v. 27, n.1, p.100109, 2007.

MAROUELLI, W. A. et al. Development and quality of peas under different soil water tensions. Pesquisa agropecuària brasileira, Brasilia, v.26, n.7, p.1041-1047. 1991a.

MAROUELLI, W. A.; SILVA, W. L. C.; MORETTI, C. L.. Plant development, production and quality of garlic bulbs under conditions of soil water deficiency. **Horticultura Brasileira**, Brasilia, v. 20, n. 3. 2002b.

MAROUELLI, W. A.; SILVA, W. L. C.; MORETTI, C. L. Response of processing tomato plants to soil water tensions under drip irrigation. **Engenharia Agricola**, Jaboticabal, v.23, n.1, p.1-8. 2003c.

MARQUES, P. A. A. COELHO, C. R. D. Estudo da viabilidade econômica da irrigaçâo da pupunheira (*Bactris Gasipaes* H.B.K.) para Ilha Solteira - SP, Brasil. **Ciência Rural**, Santa Maria, v. 33, n. 2, p. 291-297, 2003.

MARTINS, S.R.; GONZALEZ, J.F. Evapotranspiration and physiological responses of snap beans in substrate in a greenhouse with a ventilation/heating system.

Revista Brasileira de Agrometeorologia, v.3, p.31-37. 1995.

MARTINS, S. R. Desafios da plasticultura brasileira: limites socioeconômicos e tecnológicos frente às novas e crescentes exigências. **Horticultura Brasileira**, Brasilia, v. 14, n. 1, p. 133-138, May 1996.

MONTEIRO ROC; COLARES DS; COSTA R. N. T.; LEÂO M. C. S.;

AGUIAR J. V. Response function of the melon tree to different irrigation rates and nitrogen doses. **Horticultura Brasileira**, Brasilia, DF, v. 24, p. 455-459, 2006.

NASCIMENTO, M. N. et al . Plant biochemical and morphological changes in coffee buds associated with flowering events in response to meteorological elements. **Ciencia Rural**, Santa Maria, v. 38, n. 5, 2008 .

NORONHA, J. F. Production theory applied to the economic analysis of experiments. In: CONTINI, E.; ARAÙJO, J. de; OLIVEIRA, A. J.;

GARRIDO, W. E. **Farm planning**; decision models. Brasilia, EMBRAPA, 1984. 300p.

OFORI,K.; KLOGO, P.Y. Optimum Time for Harvesting Yardlong Bean *(Vigna sesquipedalis)* for High Yield and Quality of Pods and Seeds. **Journal of agriculture and social sciences**, [S. l.], v. 1, n. 2, p. 86-88, 2005.

OLIVEIRA, C. A. S. Determination of water tension in agricultural soil using a heat dissipation sensor. **Pesquisa agropecuària brasileira**, Brasilia, v. 34, n. 8. 1999.

OLIVEIRA, F. A. de; CAMPOS, T. G. S.; OLIVEIRA, B. C.. Effect of soil water tensions on the yield of herbaceous cotton. **Pesquisa agropecuâria brasileira**, Brasilia, v. 34, n. 10. 1999.

OLIVEIRA, E. C. et al. Technical and economic feasibility of pea production (*Pisum sativum* L.) grown under different irrigation rates. **Engenharia Agricola**, Jaboticabal, v. 31, n. 2, p. 324-333, 2011.

PAVINATO, P. S et al. Nitrogen and potassium in irrigated corn: technical and economic analysis of fertilization. **Ciência Rural**, Santa Maria, v.38, n. 2, p. 358-364, 2008.

PARIZI, A. R. C. et al . Irrigation levels for Kalanchoe grown in a protected environment. **Ciencia Rural**, Santa Maria, v. 40, n. 4, 2010.

PIMENTEL, L. D. et al. Economic feasibility study on the cultivation of macadamia nuts in Brazil. **Revista Brasileira de Fruticultura**, Jaboticabal - SP, v. 29, n. 3, p. 500-507, 2007.

PIMENTEL, A. A. M. P. **Olericultura no tròpico ùmido: hortaliças na Amazônia**. Sâo Paulo: Agronômica Ceres, 1985. 322p.: il.

PIRES, R. C. M et al. Crop water requirements and irrigation management. In: **Irrigation**. Pires R. C. M.; MIRANDA, J. H. (Editors). Piracicaba: FUNEP, 2002a. 410p. (Agricultural Engineering Series, 1).

PIRES, R. C. M. et al. Strawberry production under different water levels, soil cover and growing environment. **Irriga**, Botucatu, v.12, n. 3, p. 338-353. 2007b.

PONCIANO, N. J.; SOUZA, P. M.; GOLYNSKI, A. Economic evaluation of the production of pineapple (*Ananas cosmosus* L.) cultivar perola in the Norte Fluminense region. **Caatinga**, Mossoró, v. 19, n. 1, p. 82-91, 2006.

PREVEDELLO, C. L. **Fisica do solo com problemas resolvidos**. Curitiba: Autonomous Society for Advanced Studies in Soil Physics, 1996. 446 p.

PHENE, C. J.; HOWELL, T. A. Soil sensor control of high-frequency irrigation systems. **Transactions of the ASAE**, v. 27, n. 2, p. 392-396. 1984.

QUEIROZ, J. E. et al. Optimal irrigation strategies for beans: land as a limiting factor for production. **Pesquisa agropecuària brasileira**, Brasilia, v. 31, n. 1, p. 55-61, 1996.

REICHARDT, K.; TIMM, L. C.. **Soil, plant and atmosphere:** concepts, processes and applications. 1. ed. Barueri: Manole, 2004. 478p.

REIS, L. S.; SOUZA, J. L. de; AZEVEDO, C A. V. de. Evapotranspiration and crop coefficient of kaki tomatoes grown in a protected environment. V. de. Evapotranspiration and crop coefficient of kaki tomatoes grown in a protected environment. **Revista brasileira de engenharia agricola e ambiental**, Campina Grande, v.13, n.3. 2009.

RESMI, R.; GOPALAKRISHNAN, T.R.. Effect of plant growth regulators on the performance of yard long bean (*Vigna unguiculata* var. *sesquipedalis* (L.) Verdcourt). **Journal of Tropical Agriculture**, [S. l], v. 42, n. 1, p. 55-57, 2004.

ROMANINI, Carlos E. B. et al . Development and simulation of an advanced environmental control system in protected cultivation. **Revista brasileira de engenharia agricola e ambiental**, Campina Grande, v. 14, n. 11. 2010.

SA, N. S. A. de et al . Tomato crop behavior under different soil water tensions in a protected environment. **Revista brasileira de engenharia agricola e ambiental**, Campina Grande, v. 9, n. 3. 2005.

SANTANA, M. J. et al . Yield of chili peppers (Capsicum annuum L.) irrigated under different soil water tensions and doses of calcium. **Ciência e agrotecnologia**, Lavras, v. 28, n. 6. 2004 .

SANTANA, M. J. et al. Technical and economic feasibility of applying water to common bean crops (*Phaseolus vulgaris* L.). **Ciência e Agrotecnologia**, Lavras, v.33, p.532-538, 2009.

SANTOS, S. R.; PEREIRA, G. M. Behavior of American-type lettuce under different soil water tensions in a protected environment. **Engenharia Agricola**, Jaboticabal, v. 24, n. 3, p. 569- 577, Sep./Dec. 2004.

SGANZERLA, E. **Nova Agricultura:** a fascinante arte de cultivar com os plàsticos. 6. ed. Guaiba: Agropecuària, 1997. 342 p.

SILVA, G. S. Galls on meter bean stems caused by Meloydogine incognita. **Nematologia Brasileira**, v. 27, n. 2, p. 227-228. 2003.

SILVA, W. L. C.; VIEIRA, J. V.; CARRIJO, O. A. Effects of different soil moisture tensions on the carrot crop during active root system development. **Pesquisa agropecuària brasileira**, Brasilia, v. 17, n. 2, p. 211-214. 1982.

SOUSA, E. F.; BERNARDO, S.; CARVALHO, J. A. Sugarcane yield function in relation to water for three varieties in campos dos goytacazes, RJ. Engenharia Agricola, Jaboticabal, v.19, n.1, p.28-42, 1999.

SOUSA, V. F. de et al. Water use efficiency by yellow passion fruit under different irrigation levels and potassium doses. **Revista brasileira de engenharia agricola e ambiental**, Campina Grande, v. 9, n. 3. 2005b.

SOUSA, V. F. et al . Soil moisture in the root zone of passion fruit grown under localized irrigation. **Engenharia Agricola**, Jaboticabal, v. 26, n. 2. 2006a.

SOUZA, O. P. de et al . Vine quality and pineapple productivity at different planting densities and irrigation rates. **Pesquisa agropecuâria brasileira**, Brasilia, v. 44, n. 5, May 2009.

STONE, L. F. et al. Effects of soil water tension on bean productivity and growth.

Pesquisa agropecuâria brasileira, Brasilia, 1988a, v.34, n.2, p. 227-234.

STONE, L. F. et al. Effects of water suppression at different growth stages on irrigated rice production. **Pesquisa agropecuâria brasileira,** Brasilia, 1979b, v.14, n.2, p. 105-109.

UNITED STATES AGENCY FOR INTERNATIONAL DEVELOPMENT. Bora (Yard long bean) Postharvest Care and Market Preparation. [S.I.], 2004. 17 p. il. Color (Technical Bulletin n. 20). Available at: <http://pdf.usaid.gov/pdf_docs/PNACY836.pdf> Accessed on September 20, 2007.

VILAS BOAS, R. C.; CARVALHO, J. A.; GOMES, L. A. A.; SOUZA, A. M.

G. de; RODRIGUES, R. C.; SOUZA, K. J. Technical and economic evaluation of the production of two cultivars of curly lettuce as a function of irrigation rates. **Ciência e agrotecnologia**, Lavras, v. 32, n. 2. 2008b.

PART TWO - ARTICLES

ARTICLE 1

IRRIGATION MANAGEMENT FOR METER BEANS IN THE VEGETATIVE AND PRODUCTIVE PHASES IN A PROTECTED ENVIRONMENT

Prepared according to the norms of the Brazilian Journal of Agricultural and Environmental Engineering (preliminary version).

IRRIGATION MANAGEMENT FOR METER BEANS IN THE VEGETATIVE AND PRODUCTIVE PHASES IN A PROTECTED ENVIRONMENT

SUMMARY

Plant development and production are related to genetic factors, which are affected by crop management. Proper management of irrigation water requires knowledge of these procedures for each crop grown, and in particular in the phenological phases of its production cycle, since many species have critical periods during which water stress causes serious reductions in production. The aim of this study was to evaluate the effect of five levels of soil water potential on the growth and production of meter beans. The experiment was conducted in the experimental area of the Engineering Department of the Federal University of Lavras (UFLA) - Lavras-MG. The experimental design was randomized blocks, in a 5x2 factorial scheme. The treatments consisted of five soil water potentials to indicate when to irrigate (-15, -35, -55, -75 and -95 kPa), in two phenological phases (vegetative and reproductive), monitored by sensors installed at 0.15 m depth. The results showed that the most sensitive phase to water deficit was the reproductive phase, and that irrigation carried out at a matrix potential of around -15 kPa showed the best response from the crop in terms of development, production and pod quality.

Keywords: Greenhouses. Matric potential. Growth. Production. Dry matter.

1 INTRODUCTION

The growing demand for quality vegetables, coupled with problems such as the scarcity of arable land close to large consumer centers, excessive or insufficient rainfall, rational use of water, soil salinization and phytosanitary problems, as well as the globalization of the market, mean that agricultural production is becoming more and more technologically advanced in order to be competitive and serve the consumer market.

Due to these factors, alternatives have been sought to ensure the production of a large number of agricultural crops of economic interest and guarantee a year-round supply, among them irrigation and cultivation in a protected environment.

Irrigation management represents the procedures used to irrigate. For these procedures to be carried out correctly, it is necessary not only to have the means to help determine when to irrigate and how much water to apply, but also to know the response of the crops to the irrigation water management procedures adopted (FRIZZONE, 2007).

Information on the effect of irrigation water management on crops is important for: knowing the critical periods during which water deficiency causes serious reductions in production; determining when to apply water in order to maximize the efficiency of its use, of electrical energy, of the irrigation system; and maintaining favorable soil moisture conditions to guarantee production and return on investment.

Although it seems obvious that a crop's production increases with soil moisture kept close to field capacity throughout its production cycle, this relationship is not true for all crops or even phenological stages, as shown by research carried out on onion crops (ABREU et al., 1980), peas (MAROUELLI et al., 1991), eggplants (BILIBIO et al., 2010) and others. The results of the research carried out by these authors show that general recommendations for irrigation water management procedures for species in the same agronomic group, such as vegetables, for example, are inadequate in view of the different responses of species to water deficit.

The subway bean (*Vigna unguiculata* spp. *sesquipedalis* (L.) Verdc.) is a vegetable widely cultivated in the world. It originated in Central Africa (SARUTAYOPHA et al., 2007) and was first cultivated in Southeast Asia (HA et al., 2010). In Brazil, this crop was introduced in the Northeast. Later, it spread to other regions of the country, mainly to the North and Center-West (CARDOSO, 1997). In the North, it is widely consumed, mainly because it replaces the "pod" in various dishes (SILVA, 2003).

Demand in the northern region is mainly met by small producers who have this vegetable as a significant source of income. In the state of Amazonas, where consumption is widespread, plantations by farmers in the vàrzea area predominate (CARDOSO, 1997).

Despite representing yet another alternative source of income for rural producers in Amazonas, the meter bean is still produced seasonally, as both excessive rainfall and low soil humidity are unfavorable to its cultivation; as a result, its supply does not meet the demand of the regional market throughout the year (CARDOSO, 1997).

Given its economic importance and the scarcity of information on this species, the aim of this study was to obtain parameters for managing irrigation water for the crop by evaluating the effects of matrix potentials on crop development and production.

2 MATERIAL AND METHODS

The experiment was carried out in an arched greenhouse in the experimental area belonging to the Engineering Department of the Federal University of Lavras (UFLA), in Lavras, Minas Gerais. Lavras is located between the geographical coordinates: 21°14' South latitude, 45°00' West longitude and an altitude of 918m.

The region's climate, according to the Koppen classification, is humid temperate (Cwa). The average temperature of the hottest month is 22.8° C, that of the coldest month is 17.1°C, and the annual average is 20.4°C (DANTAS et al., 2007).

The soil in the experimental area was classified as Latossolo Vermelho Distroférrico, whose physical and chemical characteristics for the 0 to 30 cm layer can be seen in Table 1.

Table 1 Physical and chemical characteristics of the soil in the experiment area, for the 0 to 30 cm depth layer*.

Physical analysis

Areia	Silte	Argila	Textura	ρ_g	ρ_p	θ_{cc}	θ_{pm}
--------- (dag kg^{-1}) ---------				---- (g cm^{-3}) ----		---- (%) ----	
12,1	15,9	72,0	Argilosa	1,01	2,64	0,42	0,27

Chemical analysis

pH	MO	P	K	Ca^{2+}	Mg^{2+}	V	m	B	Cu	Fe	Mn	Zn
H_2O	g kg^{-1}	mg dm^{-3}		$cmol_c$ dm^{-3}		%		---------- mg dm^{-3} ----------				
5,8	2,3	12,9	225	4,2	1,1	71,1	0,0	1,3	8,0	31,1	56,4	16,4

V = base saturation index, m = aluminum saturation index.

* Analyses carried out at the Soil Physics and Fertility Laboratory of the DCS/UFLA.

The characteristic soil water retention curve was obtained experimentally. The values obtained were adjusted to the VAN GENUCHTEN model (1980). With the help of the SWRC 3.0 (Soil Water Retention Curve) computer program, developed by DOURADO NETO et al. (2000), the following equation was obtained, with an adjustment coefficient of 0.97:

$$\theta = 0{,}221 + \frac{(0{,}750 - 0{,}221}{\left[1 + \left(0{,}560|\psi|\right)^{1{,}548}\right]^{0{,}354}} \quad (1)$$

in which:

θ - soil moisture (cm^3 cm^{-3});

ψ - soil water potential (kPa).

The experimental design was randomized blocks, with four replications. The treatments, arranged in a 5x2 factorial arrangement, resulted from the combination of five levels of matric potential (-15, -35, -55, -75 and -95 kPa), at two stages of crop development (vegetative and reproductive). Each experimental unit consisted of a bed with a two-meter-long planting line and a density of six plants per meter. The 10 central plants were considered useful.

The vegetative phase corresponded to the period between emergence and the time when 50% of the flowers on the plants in the -15 kPa treatment were open (50 days after emergence, DAE) and the reproductive phase between the beginning of anthesis and the end of the harvest (82 DAE), when irrigation was stopped.

The beds were internally delimited by black polyethylene film (0.03 mm), in order to prevent water from moving between them and not interfering with the treatments. The beds were built 2.0 m long, 0.40 m wide and 0.30 m deep for a total of 240 dm^3 , spaced 0.60 m apart lengthways and 0.40 m apart crossways. The soil used in the beds was air-dried, sieved through a 4 mm mesh sieve, homogenized and taken back to the bed.

The planting and top dressing fertilizations were carried out according to the soil fertility analysis and the recommendations for growing beans (CFSEMG, 1999). 150 kg ha^{-1} of N and 230 kg ha^{-1} of P2O5 were used. The planting fertilizer was made during soil preparation, adding all the P2O5 content. Part of the N fertilization was done on the day of sowing (1/3), and the rest (2/3) was divided into two applications, carried out at 25 and 40 DAE. The nutrient sources used were urea and triple superphosphate.

The variety used was "brown seed". 36 seeds were sown per bed, distributed in 12 planting furrows at a depth of 1.0 cm (1/10/2009). The seedlings emerged completely on 05/10/2009 (when 100% of the seedlings had emerged), the date on which the treatments began to be differentiated. Three days after emergence (08/10/2009), 12 plants were thinned out (using the espalier system, with the last wire at a height of 1.8 m above the ground).

Drip irrigation was carried out using six self-compensating drippers in each bed, spaced 0.33 m apart, with a flow rate of 4 L h^{-1} and a working pressure of 100 kPa. The average uniformity of water application was 98.8%, obtained shortly after the irrigation system was set up.

Irrigation was managed based on the average reading of the matric potential observed in the sensors installed in three experimental units of each treatment. Tensiometers were used to monitor the treatments with a matric potential equal to or lower than -75 kPa, while the other treatments used granular matrix sensors (GMS). Once the time for irrigation had been determined, a sufficient amount of water was applied for the soil moisture to reach field capacity. The effective depth of the root system was 0.30 m, with the moisture sensors installed at half this depth (0.15 m). The readings were taken daily (at 9, 12, 15 and 17 hours) using a digital tensimeter with a sensitivity of 0.1 kPa.

Moisture at field capacity was obtained through an *on-site* test, equivalent to a matric potential of -7 kPa. The applied blade was calculated taking into account: the estimated average matric potential (for each treatment); the soil water retention equation and the relative humidity at field capacity.

The pods were harvested 12 days after anthesis. This criterion was based on observing a change in the color of the pods (when they turned from an intense dark green to a bright light green), which was considered the ideal point for harvesting. To standardize this procedure, the flowers that were open were identified daily in the morning with different coloured threads.

The effects of the treatments were evaluated on neck diameter, shoot height, leaf dry matter, branch dry matter, shoot dry matter, fresh pod mass, pod length, number of pods per plant and yield of the metro bean crop.

The neck diameter, height and dry matter of the aerial part were determined accordingly. Plant height was obtained at the end of the vegetative cycle and neck diameter, plant height and plant dry matter at the end of the reproductive cycle. All the development and production characteristics were obtained from the 10 central plants in the experimental unit.

The data collected was subjected to analysis of variance, regression analysis and Tukey's test to compare means. In the regression analysis, the linear and 2nd degree polynomial models were tested. The regression equations were chosen based on the significance of the regression coefficients, at 1% and 5% probability, by the F test, and on the highest value of the coefficient of determination R^2 . The statistical tests were carried out using the SISVAR statistical program version 5.0 (Ferreira, 2008).

3 RESULTS AND DISCUSSION

As can be seen in Table 2, the average values of the matric potential used to monitor irrigation in the different treatments were different from those initially established, although they were close. These results are due to the fact that the readings were not taken by automated systems, which made it impossible to match the pre-established values (FIGUERÊDO et al., 2008).

Table 2 also shows that the amount of water applied and the number of irrigations increased as the matric potential increased (soil with more easily available water), which was expected. According to TAIZ and ZAIGER (2009), plants grown in soil with more easily available water show greater development, leaf area and transpiration and, consequently, require more frequent irrigations and a greater volume of water to replace what has been used for metabolism and also "given up" to the atmosphere.

Table 2 Water applied and number of irrigations carried out at each phenological stage of the meter bean as a function of matric potential

VE (kPa)	-15		-35		-55		-75		-95	
VO (kPa)	-16		-37		-53		-76		-92	
	L	N	L	N	L	N	L	N	L	N
FV	141,5	15	128,7	11	128,9	9	76,0	3	27,3	2
FR	150,9	16	145,2	12	126.9	11	66,4	5	54,0	4

VE = established value; VO= average value obtained; FV= vegetative phase; FR= reproductive phase; L= Làmina; N= number of irrigations

The analysis of variance (Table 3) showed a significant effect of matric potential (T), phases (F) and the T x F interaction for the variables diameter (D), leaf dry mass (MSF), branch dry mass (MSR) and aerial part dry mass (MSPA), with no response found for height ($p > 0.05$). The growth behavior of the meter bean in height is consistent with that described by Oliveira (2009), who found no significant effect on the height of cucumber plants with the reduction of the matrix potential (-15, -30, -60 and -120 kPa) in the vegetative phase. This is also consistent with the results obtained by Sà (2004) for tomato plants, irrigated with decreasing levels of matric potential (-15, -30, -50, -70, -120 and -170 kPa).

Table 3 Summary of the analysis of variance for diameter, height, leaf dry matter (LSM), branch dry matter (BDM) and aerial part dry matter (APDM) of meter beans under different levels of matric potential (T) and phenological phases (F).

Sources of variation	GL	Mean square				
		Diameter	Height	MSF	MSR	MSPA
T	4	3,60*	2030,16^{ns}	5538,15*	3098,20*	174,63*
F	1	0,33*	1834,15^{ns}	2468,83*	17351,39*	6860,47*
T x F	4	0,19*	472,40^{ns}	304,92*	3018,79*	1869,41*
Block	3	0,08*	187,70^{ns}	100,38*	1606,90*	1345,65*
Waste	27	0,07	0.09	55,73	6378,13	517,82
CV (%)		3,59	12.4	11,62	10,87	11,08

*, and^{ns} - Significant at 5% and not significant, respectively.

As can be seen in Graph 1, the plants that grew in the wettest soil (-15 kPa) in the FV had higher D, MSF, MSR and MSPA than those that grew in drier soil (-35, -55, -75 and -95 kPa). However, under the same growing conditions at FR, the plants behaved differently in terms of MSR compared to the other variables, because even when grown in drier soil at this stage, the average MSR value remained approximately constant. One possible explanation for this behavior is the fact that most of the branches developed during the PV, when the plants were irrigated at the highest matric potential (-15 kPa); and as the branches do not detach from the plants under unfavorable environmental conditions, it was to be expected that the RMS would not be reduced under water deficit in the FR.

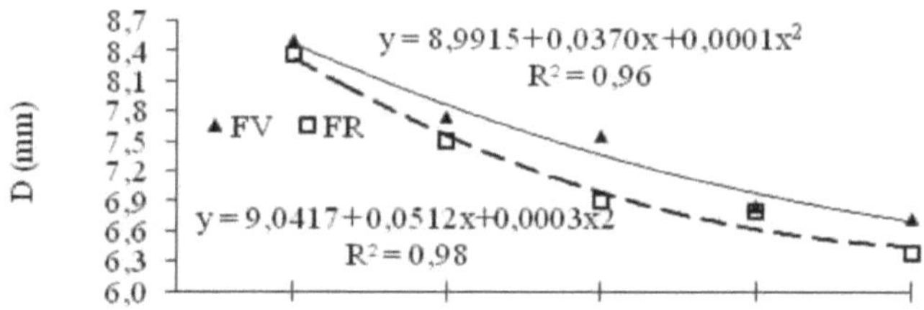

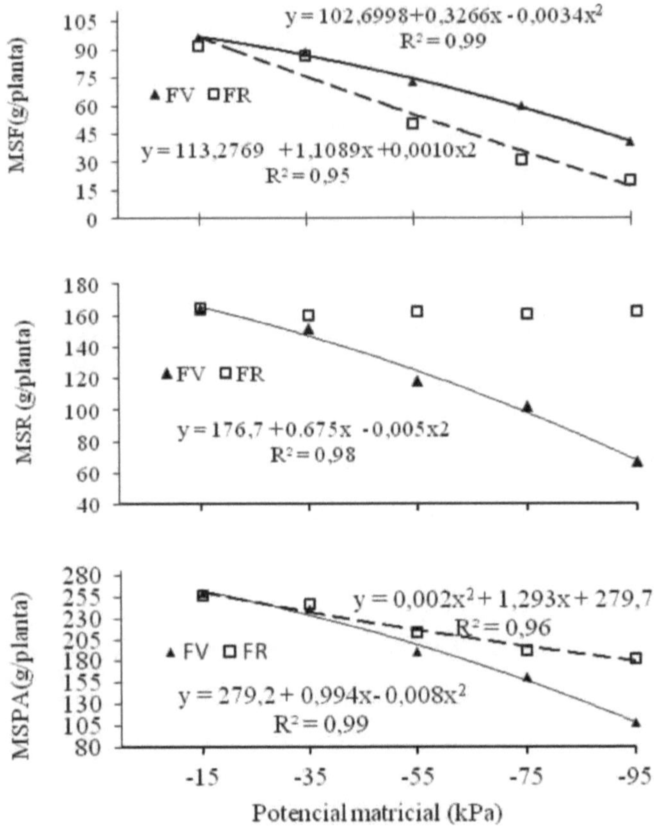

Graph 1 Data on diameter (A), dry mass of leaf (B), dry mass of branch (C) and dry mass of aerial part as a function of the matric potential applied in the two phenological phases.

The adjusted regression equations indicate the superiority of plants grown under irrigation at the highest matric potential (-15 kPa). These results show that irrigation at lower matric potentials (-35, -55, -75 and -95 kPa) promotes a reduction in growth and development, as has also been observed by various researchers (SANTOS and PEREIRA, 2004; FARIAS et al., 2005; MACEDO et al., 2010).

The tendency for D, SPS, RMS and SPS production to decrease as the matric potentials decrease (Figure 1), can be attributed to the difficulty the meter bean plants had in absorbing water, since in a situation of lower matric potential the water is retained more strongly on the surface of the colloids; thus, the plants need to spend more energy to acquire water for use in their vital physiological processes and this negatively affects growth characteristics.

The results obtained in this study agree with those obtained by MAROUELLI et al. (2003a), who also found a reduction in the MSPA production of garlic plants when irrigations during the vegetative and reproductive phases were carried out at matrix potentials between -20 and -160 kPa, and also with MAROUELLI et al. (2003b) for tomato plants irrigated during the productive phase at matrix potentials between -15 and -70 kPa. These authors attributed these results to plant adaptations to water deficit in the vegetative phase, such as greater root growth, lower leaf area index and greater leaf cuticle thickness. The differences observed in the response of these plant species must be related to the matrix potential levels used and the different sensitivity of each species to water deficit in the phenological phase of its cycle, reinforcing the assertion that each cultivated species should receive differentiated irrigation management, according to its needs.

Table 4 shows the results of the mean tests for diameter (D), leaf dry matter (SDF), branch dry matter (SDF) and shoot dry matter (SPD) of the meter bean plants subjected to different irrigation water management in the vegetative (VF) and reproductive (RF) phases of the crop. It can be seen that D and MSF had higher averages when the plants were grown under water deficit in the VF. These results indicate that in the FR the plants were more sensitive to irrigation water management carried out at lower soil moisture levels (equivalent to lower matric potential levels), as higher D and MSF values indicate more robust plants vegetatively and with greater photosynthetic capacity, which is essential for plant production.

Table 4 Average data for diameter, leaf dry matter (LSM), branch dry matter (BDM) and aerial dry matter (ASM) of the subway bean as a function of the treatments applied in the vegetative phase (VF) and reproductive phase (RF).

Phase	Diameter (mm)	MSF (g/plant)	MSR (g/plant)	MSPA (g/plant)
FV	7,47 a	71,73 a	120,64 b	192,29 b
FR	7,19 b	56,27 b	162,26 a	218,48 a

Means followed by the same letter vertically do not differ at 5% probability by Tukey's test.

Graph 2 shows that the reduction in SFA of plants irrigated at lower matrix potentials was similar in the two phenological phases, but the adaptation mechanism used by the plants under these growing conditions was different. While the reduction in SFA of plants grown under water deficit in PV was due to fewer leaves emerging and a reduction in the area of expanding leaves, in FR the reduction in SFA was due to the fall of fully expanded leaves, as observed during the experiment.

Higher MSPA values indicate better plant development, however, according to the results obtained in this study, this parameter was not indicative of greater plant growth, because, as can be seen in Table 4, even with a higher average MSPA, the meter bean plants subjected to water deficit in the FR showed less growth in diameter, a fact that reinforces this claim.

As shown in Table 4, the greatest contribution to MSPA production, regardless of the stage at which the plants were subjected to water deficit, was due to MSR, as its average was higher than that provided by MSF. However, it can be seen that SPS was higher in the PV (Table 4). This reinforces the assertion that SBM contributed the most to the plants' best performance, since plants with greater SBM have greater leaf area and therefore greater potential to synthesize the metabolites essential for crop development.

The analysis of variance (Table 5) indicated a significant effect of the treatments, phases and the interaction between them for number of pods (NF), pod mass (MF), pod length (CF) and pod yield (RF).

Table 5 Summary of analysis of variance for number of pods (NF), pod mass (MF), pod length (CF) and pod yield (RF) of meter beans

Sources of variation	GL	Mean square			
		NF	MF	CF	RF
Treatment (T)	4	997.70*	2.93*	18.74*	179602.99*
Phase (F)	1	36.48*	39.20*	132.13*	100761.44*
T x F	4	3.74*	2.55*	11.69*	6032.65*
Block	3	0.87*	0.13*	3.33*	873.95*
Waste	27	4.26	0.58	2.15	1810.67
CV (%)		4,7	6,7	3,6	8,4

* and^{ns} - Significant at 5% and not significant, respectively.

Graph 2 shows that NF and RF decreased with the matrix potential applied, regardless of the phenological stage.

However, this behavior was only observed for MF and CF when the water deficit was applied to FR. Based on these results, it can be inferred that irrigation below -15 kPa in the FV influenced production quantitatively by reducing the NF, while in the FR, as well as affecting the number of pods, it also affected the quality of the pods by reducing the MF.

The highest estimated values for NF, MF, CF and RF are obtained when irrigation is carried out at the highest matric potential (-15 kPa) (Graph 2). This result suggests that in order to obtain better yields and better quality pods, meter bean irrigation should be carried out more frequently and, therefore, with the soil closer to field capacity.

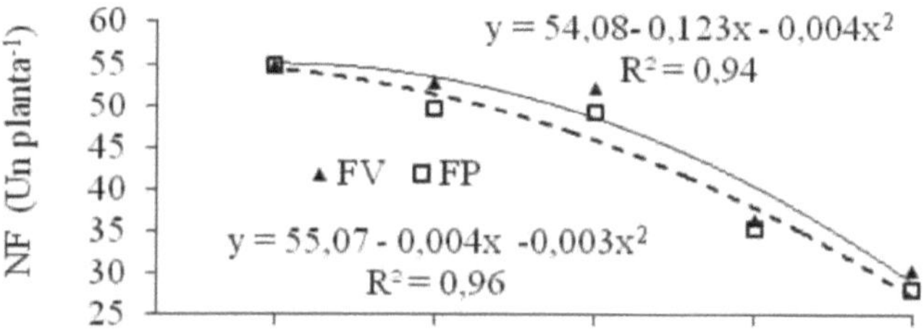

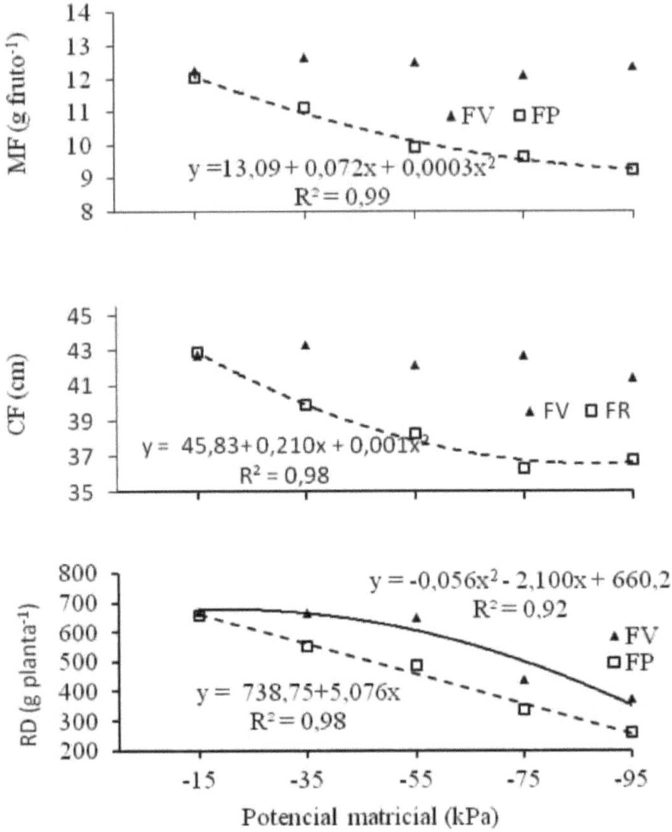

Graph 2 Data on number of pods (A), pod mass (B), pod length (C), and yield (D) as a function of the matric potential applied in the two phenological phases.

The effect of deficit irrigation on RF was more pronounced than that observed in PV (Table 6), suggesting that irrigation water management with a deficit in RF limits the productive potential of meter bean plants to a much greater extent.

Table 6 Mean test for number of pods (NF), pod mass (MF), pod length (CF) and pod yield (RF) of meter beans for the vegetative phase (FV) and reproductive phase (FR).

Phase	NF (one plant $)^{-1}$	MF (g plant $)^{-1}$	CF (cm pod $)^{-1}$	RF (g plant $)^{-1}$
FV	45,0 a	12,38 a	42.48 a	559.91a
FR	44,0 b	10,39 b	38.84 b	459.53 b

Means followed by the same letter vertically do not differ at 5% probability by Tukey's test.

In natural and cultivated ecosystems, the photosynthetic activity of plants results in the conversion of solar energy into chemical energy, which is essential for plant production. It can therefore be said that plant production is dependent on photosynthesis. Therefore, the superior yield obtained by the plants irrigated with water deficit in the PV was achieved due to their greater potential for producing photoassimilates, as there was possibly greater interception of solar radiation by the leaves as a result of the greater production of SPS (Table 4).

4 CONCLUSION

The greatest plant growth, yield and product quality were observed in treatments irrigated with a matric potential of -15 kPa.

The meter bean was more sensitive to water deficit in the reproductive phase, in which the yield and quality of the pods were lower than in the vegetative phase.

REFERENCES

ABREU, T. A. S.; SILVEIRA, P. S.; PEREIRA, G. T. Analysis of onion production under different irrigation regimes. **Pesquisa agropecuària brasileira**, Brasilia, v. 15, n. 2, p. 233-236, 1980.

CARDOSO, M.O. **Unconventional vegetables from the Amazon**. EMB RAPA- MA. Brasilia - DF, 1997.

BILIBIO, C.; CARVALHO, J. A.; MARTINS, M.; REZENDE, F. C.;

FREITAS, E. A.; GOMES, L. A. A. Vegetative and productive development of eggplant subjected to different soil water tensions. **Revista Brasileira de Engenharia Agricola e Ambiental**, v. 14, n. 7, p. 730-735, 2010.

SOIL FERTILITY COMMISSION OF THE STATE OF MINAS

GERAIS. **Recommendations for the use of correctives and fertilizers in Minas Gerais**. 5. ed. Lavras, 1999. 359 p.

DANTAs, A. A. A.; CARVALHo, L. G. de; FERREIRA, E. Climate classification and trends in Lavras, MG. **Ciência e agrotecnologia**, v. 31, n. 6, p. 1862-1866, 2007.

DoURADo NETo, D. ; NIELsEN, D. R. ; HoPMANs, J. W. ; REICHARDT, K. ; BACCHI, o. o. s. software to model soil water retention curves (sWRC, version 2.00). **Scientia Agricola**, Piracicaba, v. 57, n.1, p. 191-192, 2000.

FARIAs, M.F. de; SAAD, Joâo Carlos C.. Growth and quality of chrysanthemums grown in pots in a protected environment. **Horticicultura Brasileira**, Brasilia, v. 23, n. 3, p. 740-742, 2005.

FERREIRA, D. F. SISVAR: a program for statistical analysis and teaching.

Revista Symposium, Lavras, v. 6, p. 36-41, 2008.

FIGUERÊDO, S. F. de ; POZZEBON, É. J. ; FRIZZONE, J. A.; AZEVEDO, J. A. ; GUERRA, A. F. ; SILVA, EUZÉBIO, M. da . Bean irrigation management based on technical and economic criteria in the cerrado. **Irriga**, v. 13, p. 378-391, 2008.

FRIZZONE, J. A. Irrigation planning using optimization techniques. **Revista Brasileira de Agricultura Irrigada**, Fortaleza, v. 1, n. 1, p. 24-49, 2007.

HA, T. J. et al. Identification and Characterization of Anthocyanins in Yard- Long Beans (Vigna unguiculata ssp. sesquipedalis L.) by High-Performance Liquid Chromatography with Diode Array Detection and Electrospray Ionization/Mass Spectrometry (HPLC-DAD-ESI/MS) Analysis. **Journal of Agricultural and Food Chemistry**, [S. l], v. 58, n. 4, p. 2571-2576. 2010. 2010.

VAN GENUCHTEN, M.Th. A closed form equation for predicting the hydraulic conductivity of unsaturated soils. **Soil Science Society of American Journal**, Madison, v. 41, p. 892-8, 1980.

MACEDO et al.Performance of an automated irrigation system through soil water tension. **Revista Brasileira de Agricultura Irrigada**, Fortaleza, v. 4, n. 2, p. 78-81, 2010.

MAROUELLI, W. A. et al. Development and quality of peas under different soil water tensions. **Pesquisa agropecuària brasileira**, Brasilia, v. 26, n. 7, p. 1041-1047, 1991.

MAROUELLI, W. A.; SILVA, W. L. C.; MORETTI, C. L. Response of processing tomato plants to soil water tensions under drip irrigation. **Engenharia Agricola**, Jaboticabal, v. 23, n. 1, p. 1-8, 2003.

MAROUELLI, W. A.; SILVA, W. L. C.; MORETTI, C. L. Plant development, production and quality of garlic bulbs under conditions of soil water deficiency. **Horticultura Brasileira**, Brasilia, v. 20, n. 3, p. 470-473, 2003.

OLIVEIRA, E. C. Irrigation management of the Japanese *cucumber* crop (*Cucumis sativus* L.) in a protected environment. 2009. 88 p. Dissertation (Master's Degree in Agricultural Engineering) - Federal University of Lavras, Lavras, MG.

SA, N. S. A. Growing tomato plants (*Licopersicon esculentum* Mill.) in a protected environment under different soil water tensions. 2004. 71 p. Dissertation (Master's Degree in Irrigation and Drainage) - Federal University of Lavras, Lavras, MG.

SARUTAYOPHA,T.; NUALSRI, C.; SANTIPRACHA, Q.; SAEREEPRARAT, V. Characterization and genetic correlations among 37 yardlong bean and cowpea accessions based on morphological characters and RAPD analysis.

Sangklanakarin Journal of Science and Technology, [S. l], v. 29, n. 3. p. 591-600, 2007.

SANTOS, S. R.; PEREIRA, G. M. Behavior of American-type lettuce under different soil water tensions in a protected environment. **Engenharia Agricola**, Jaboticabal, v. 24, n. 3, p. 569- 577, 2004.

SILVA, G. S. Galls on meter bean stems caused by Meloydogine incognita. **Nematologia Brasileira**, v. 27, n. 2, p. 227-228, 2003.

TAIZ, Lincoln; ZEIGER, Eduardo. **Plant physiology.** 4. ed. Porto Alegre: Artmed, 2009. 819 p.

ARTICLE 2

TECHNICAL AND ECONOMIC ANALYSIS OF IRRIGATION FOR METRO BEANS IN A PROTECTED ENVIRONMENT

Prepared according to the rules of the journal Engenharia Agricola (preliminary version)

TECHNICAL AND ECONOMIC ANALYSIS OF IRRIGATION FOR METRO BEANS IN A PROTECTED ENVIRONMENT

SUMMARY

The experiment conducted in Lavras-MG, in a protected environment, aimed to estimate the water levels that maximize productivity and economic return in the cultivation of metre beans and to analyse the economic viability of these irrigation management strategies. The experimental design was randomized blocks, with five treatments and four replications. The treatments consisted of five irrigation rates: 40, 70, 100, 130 and 160 % of the water replenishment rate up to field capacity. The water levels that maximize productivity and economic return were obtained from the regression model fitted to the productivity data and the product price and water cost ratios. Economic viability was obtained based on the benefit/cost ratio. It was concluded that: the maximum yield of pods, 35,160.6 kg ha^{-1} , was estimated with a water level of 434.4 mm and the level of maximum economic return was estimated at 431.3 mm, with a yield of 35,156.6 kh ha^{-1} ; that it is economically viable to grow meter beans under these growing conditions; and that the expected return is R$ 1.7 for every real invested.

Keywords: *Vigna unguiculata* spp . *Sesquipedalis*. Irrigation management. Economic viability. Benefit/cost ratio.

1 INTRODUCTION

The meter bean (*Vigna unguiculata* spp. *sesquipedalis*, L.) is an annual, climbing plant with three-lobed leaves and large violet-blue flowers. It produces long pods (0.5 - 1.0 m) that are harvested at an immature stage (XU et al., 2011). The crop is tolerant of high temperatures and acidic soils (FLYMAN and AFOLAYAN, 2008).

This vegetable is an important source of minerals and protein in the diet of Southeast Asia, West India and China (CHEN et al., 2007; FLYMAN and AFOLAYAN, 2008). In China, the annual production of this vegetable in a regular year is approximately 7.25 million tons (ZHU et al., 2007).

In Brazil, especially in the north, this species is of great social and economic importance, as it is grown by small farmers, who produce both for consumption and on a commercial scale (CARDOSO, 1997). According to this author, the production of this vegetable is seasonal, since excess rainfall is harmful to the crop and production is concentrated mainly at the end of the rainy season, when producers take advantage of the natural humidity of the lowland areas when the rivers begin to flow.

In view of this fact, it is necessary to research alternatives that, in addition to enabling a stable supply of agricultural products, promote income generation, among which irrigation and cultivation in a protected environment stand out (MAGGI et al., 2006; DERMITAS and AYAS, 2009).

As farmers are focused on profit, in order to make a decision on whether or not to adopt new cultivation alternatives, the financial result to be obtained from the investment of resources must be analyzed. To this end, it is essential to obtain information on the return on invested capital.

Vegetables are demanding of water. However, supplying water at levels that can meet their demand in order to express the maximum genetic potential in productive terms may not be the best management alternative, since the maximum biological productivity of the crop does not always ensure the maximum economic return. Therefore, it is necessary to define an irrigation water management strategy that meets the needs of the plants and, at the same time, ensures the maximum financial return.

The combination of production factors to adjust the yield of irrigated crops to economically adequate levels can be obtained using mathematical models to predict the productive response of plants to water application (MONTEIRO et al., 2006;

MACIEL et al., 2007; OLIVEIRA et al., 2011). These models, known as the "water-crop production function", are widely used in economic analysis studies. The use of these models is justified by the need to find an agroeconomic indicator without having to resort to complex methods that are difficult to operate.

The choice of model should be based on its ability to best represent the experimental data. Research carried out with different crops has shown that the second-degree polynomial is usually used to describe the production function in economic analyses of agricultural research (KOETZ et al., 2008a; VILAS BOAS et al., 2008; BILIBIO et 2010).

The aim of this study is to evaluate, technically and economically, the effect of different drip irrigation rates on the productivity of the meter bean crop grown in a protected environment.

2 MATERIAL AND METHODS

The experiment was carried out in a greenhouse in the experimental area belonging to the Engineering Department of the Federal University of Lavras (UFLA), in Lavras, Minas Gerais. Lavras is located between the geographical coordinates: 21°14' South latitude, 45°00' West longitude and an altitude of 918m.

The region's climate, according to the Koppen classification, is humid temperate (Cwa). The average temperature of the hottest month is 22.8° C, that of the coldest month is 17.1°C, and the annual average is 20.4°C (DANTAS et al., 2007).

The soil in the experimental area was classified as Latossolo Vermelho Distroférrico, whose physical and chemical characteristics for the 0 to 30 cm layer can be seen in Table 1.

Table 1 Physical and chemical characteristics of the soil of the experimental area, for the layer 0 - 30 cm*. **Physical and chemical characteristics of the soil of the experimental area, for the layer 0 - 30 cm***

Physical analysis

Areia	Silte	Argila	Textura	ρ_g	ρ_p	θ_{cc}	θ_{pm}
---------	(dag kg^{-1})	---------	Argilosa	---- (g cm^{-3}) ----		---- (%) ----	
10,1	16,9	73,0		1,01	2,64	0,42	0,27

Chemical analysis

pH	MO	P	K	Ca^{2+}	Mg^{2+}	V	m	B	Cu	Fe	Mn	Zn
H_2O	g kg^{-1}	mg dm^{-3}		cmol$_c$ dm^{-3}		%		----------		mg dm^{-3}	----------	
6,0	2,9	29,9	306	3,0	0,7	74,9	0,0	0,8	10,1	43,1	36,1	7,4

V = base saturation index, m = aluminum saturation index.

* Analyses carried out at the Soil Physics and Fertility Laboratory of the DCS/UFLA.

The experimentally obtained values of the retention curve were adjusted to the Van Genuchten model, using the method of minimizing the square of the deviations (VAN GENUCHTEN, 1980). With the help of the computer program Soil Water Retention Curve (SWRC 3.0), developed by DOURADO NETO et al. (2000), the following equation was obtained, with an adjustment coefficient of 0.97:

$$\theta = 0{,}221 + \left(\frac{0{,}750 - 0{,}221}{\left[1 + \left(0{,}560|\psi|\right)^{1{,}548}\right]^{0{,}354}} \right) \quad (1)$$

in which,

θ - soil moisture (cm^3 cm^{-3}), and

ψ - matric potential (kPa).

The experimental design was randomized blocks, with five treatments and four replications. The treatments were applied based on different levels of water replenishment in the soil, with Ti, T2, T3, T4 and T5 corresponding, respectively, to 40%, 70%, 100%, 130% and 160% of the control treatment (T3 - 100% water replenishment to bring the soil moisture to field capacity). Each experimental unit consisted of a bed with a planting line two meters long and a density of six plants per meter. The 10 central plants were considered useful.

The beds were internally delimited by black polyethylene film (0.03 mm), in order to prevent water from moving between them and not interfering with the treatments. The beds were built 2.0 m long, 0.40 m wide and 0.30 m deep for a total of 240 dm^3 , spaced 0.60 m apart lengthways and 0.40 m apart crossways. The soil used in the beds was air-dried, sieved through a 4 mm mesh sieve, homogenized and taken back to the bed.

The planting and top dressing fertilizations were carried out according to the soil fertility analysis and the recommendations for growing beans (CFSEMG, 1999). 150 kg ha^{-1} of N and 230 kg ha^{-1} of P_2O_5 were used. The planting fertilizer was applied during soil preparation, adding all the P content. Part of the N fertilizer was applied on the day of sowing (1/3), and the rest (2/3) was divided into two applications at 25 and 40 DAE. The nutrient sources used were urea and triple superphosphate.

The variety used was "brown seed". 36 seeds were sown per bed, distributed in 12 planting furrows at a depth of 1.0 cm (1/10/2009). The seedlings emerged completely on 05/10/2009 (when 100% of the seedlings had emerged), the date on which the treatments began to be differentiated. Three days after emergence (08/10/2009), 12 plants were thinned out (using the espalier system, with the last wire at a height of 1.8 m above the ground).

Drip irrigation was carried out using six self-compensating drippers in each bed, spaced 0.33 m apart, with a flow rate of 4 L h^{-1} and a working pressure of 100 kPa. The average uniformity of water application was 99.1%, obtained shortly after the irrigation system was set up.

Irrigation management was based on the reading of the matric potential observed in four tensiometers installed at a depth of 0.15 m in the experimental units of the treatment with 100% water replenishment in the soil. The readings were taken daily (at 9, 12, 15 and 17 o'clock) using a digital tensimeter with a sensitivity of 0.1 kPa. Irrigation was carried out when the average matric potential in the tensiometers reached -15 kPa.

Moisture at field capacity was obtained through an *on-site* test, equivalent to a matric potential of -7 kPa (equation 1). The applied rate was calculated taking into account the estimated average matric potential, the soil water retention equation and the relative humidity at field capacity. The rates applied in each experimental unit were defined according to the percentages for each treatment.

The pods were harvested 12 days after anthesis. This criterion was based on observing a change in the color of the pods (when they turned from an intense dark green to a bright light green), which was considered the ideal point for harvesting. To standardize this procedure, the flowers that were open were identified daily in the morning with different coloured threads.

The results were statistically evaluated by polynomial regression analysis to the second degree. The adjusted regression model was chosen on the basis of the significance of the regression coefficients ($P < 0.05$) and the value of the coefficient of determination R^2. The statistical tests were carried out using the statistical program SISVAR version 5.0 (FERREIRA, 2008).

The production function used for the economic analysis was obtained from the regression model fitted to the productivity data, represented by equation 2:

$$Y_e(W_i) = B_0 + B_1 W_i + B_2 W_i^2 \quad (2)$$

in which,

$Y_e(W_i)$ - estimated productivity, kg ha^{-1};

W_i - applied water line, mm, and

B_0, B_1 and B_2 - regression coefficients, assuming that $B_1 > 0$ and $B_2 < 0$.

The total cost of production (TC, R$ ha^{-1}) can be divided into fixed costs (CF, R$ ha^{-1}) and variable costs (CV, R$ ha^{-1}). Fixed costs are those that occur regardless of the number of hours the irrigation system is in operation (R$ ha^{-1}), during the analysis period considered. Equations 3, 4 and 5 were used to obtain them:

$$CT = CF + CV \tag{3}$$

$$CF = C_p + C_{fi} \tag{4}$$

$$CV = C_w . W_i \tag{5}$$

in which,

C_p - cost of crop production, R$ ha^{-1} ;

C_{fi} - fixed cost of irrigation, R$ ha^{-1} , and

C_w - cost of irrigation water, R$ mm ha^{-1} ;

The cost of production of the crop (C_p), without taking into account the cost of applying irrigation, according to the price survey on the local market was R$ 17,839.6 ha^{-1} .

In order to calculate the fixed and variable costs of irrigation, a drip irrigation system was considered to meet the following characteristics: irrigated area of 1 ha; total head of 50 m; pumping unit flow of 11.8 m^3 ha^{-1} h^{-1} ; motor pump set efficiency of 50%; and 4 hours a day^{-1} of irrigation system operation. Under these conditions, the cost of the irrigation system, according to information obtained from companies that design and sell irrigation equipment, was R$9,960.48, including all the work needed to get it up and running.

The fixed cost of irrigation (C_{fi} , R$ ha^{-1}) was calculated using the capital recovery factor (CRF), which provides a coefficient that allows the fixed cost of this investment to be calculated from the value of the investment. The C_{fi} was R$ 587.62 ha^{-1} cycle^{-1} , which was calculated using equations 6 and 7:

$$C_{fi} = \frac{FRC.S_i}{n} \tag{6}$$

$$FRC = \frac{j.(1+j)^{vu}}{(1+j)^{vu} - 1} \tag{7}$$

in which,

FRC - capital recovery factor, dimensionless;

J - annual interest rate, 0.12;

vu - useful life of the irrigation system, 10 years.

Si - cost of the irrigation system; and

n - number of crop cycles per year, 3.

Under the current water resources management model, the cost of irrigation water can be considered zero. However, for the purposes of economic analysis, this cost was obtained based on the cost of electricity consumed for pumping (C_{EE}, R$ mm^{-1}) and the cost of maintaining and operating the irrigation system (C_{mo}, R$ mm^{-1}). Assuming that small rural properties fall into consumption class B2 (BILIBIO et al., 2010), the cost of water (C_w) was calculated using equations 8, 9 and 10. Thus, Cw was 2.02 R$ mm^{-1} ha^{-1} .

$$C_w = C_{EE} + C_{mo} \quad (8)$$

$$C_{EE} = \frac{10.H.0,736.T_e.L_{R100\%}}{270.\eta} \quad (9)$$

$$C_{mo} = \frac{0,02.S_i}{n} \quad (10)$$

in which,

H - manometric height, m;

T_e - electricity tariff, 0.68 R$ kWh^{-1} ;

$L_{RIOO\%}$ - water level applied in the treatment with 100% replenishment, mm;

η - efficiency of the motor-pump set, decimal.

The water level to be applied to obtain maximum productivity (W_m) was calculated by deriving the total physical product (given by equation 2) in relation to the variable factor and equating the derivative obtained to zero. At this point, the total physical product is maximum. Thus, we have

$$W_m = -\frac{b}{2c} \quad (11)$$

The irrigation water management strategy to be adopted should maximize the economic return (profit). This point is reached when the first-order derivative of equation 2 in relation to W is equated to the ratio between the cost of the water factor (C_W) and the price of the product (P_y), giving the equation for the optimum economic rate (W*):

$$W^* = W_m + \frac{C_w}{2cP_y} \quad (12)$$

in which,

Cw - cost of water factor, R$ mm^{-1} , and

Py - product price, R$ kg^{-1} .

The net revenue (RL, R$ ha^{-1}) was obtained from the difference between the gross revenue (RB, R$ ha^{-1}) and the total cost (CT), according to the expression:

$$RL_{wi} = RB - CT \quad (13)$$

$$RB = P_y.W_i \quad (14)$$

The analysis of the economic viability of meter bean irrigation was carried out based on the benefits of its adoption and the investments and costs of its implementation, using the profitability indicator of the benefit/cost ratio (B/C). As a criterion for decision-making, the minimum value above 1 was considered, and the B/C ratio must therefore exceed this value for the activity to be considered economically viable (ARAÙJO et al., 2011). This indicator was quantified using equation 15.

$$B/C = \frac{RB}{CT} \quad (15)$$

3 RESULTS AND DISCUSSION

The total amounts of irrigation applied during the experiment in the treatments corresponding to the levels of water replenishment are shown in Table 2. In all treatments, a 75 mm layer was applied to ensure seed germination and plant establishment. The irrigation treatments were then differentiated.

Table 2 Treatments, applied water depth in the initial phase (w_{ii}), applied water depth during treatment (w_{it}) and total depth (WT) irrigation applied during the experiment. **Tratment, applied water depth in the initial phase (w_{ii}), applied water depth during treatment (w_{it}) and total depth (WT) irrigation applied during the experiment**

Làmina	Treatments (%)				
	40	70	100	130	160
w_{ii} (mm)	75	75	75	75	75
w_{it} (mm)	124,8	218,9	312,1	405,7	499,4
WT (mm)	199,8	293,9	387,1	480,7	574,4

The results of the analysis of variance (Table 3) showed a significant effect (P<0.05) for all the variables analyzed (fresh mass, number of pods and yield), indicating the dependence of the effects of the treatments used in the experiment. Similar results were obtained by OLIVEIRA et al. (2011) whose results were significant for yield and number of pods when applying 50; 75; 100; 125 and 150% of the water consumed by the pea crop (*Pisum sativum* L.).

Table 3 Summary of regression analysis of variance of data on the number (NV), fresh weight (FW) and production (DP) of pods of asparagus bean depending on the depth of water applied. **Summary of regression analysis of variance of data on the number (NV), fresh weight (FW) and production (DP) of pods of asparagus bean depending on the depth of water applied**

Sources of variation	GL -	NV	Mean square MF	PD
RL	1	271,72*	14,53*	100.549,09*
RQ	1	887,32*	0,73*	139.800,04*
Deviations	2	48,41*	0,11*	6.968,09*
Blocks	3	14,85*	0,32*	3.648,35*
Waste	12	11,56	0,07	2.093,75*
CV (%)		8,93	2,15	9,93

RL = Linear regression; RQ = Quadratic regression; * = significant at 5% probability by the F test; CV = Coefficient of variation; R^2 = coefficient of determination.

Although a decrease in crop yield is expected in response to increased water application, a significant response is not always obtained, as shown by the results of research carried out by KOETZ et al.(2006)b. These authors found no statistical effect on the yield of melon plants as a result of irrigation with water levels equivalent to 50, 75, 100 and 125% of Class "A" tank evaporation. On the other hand, SOUZA et al (2009), working with the same species, growing conditions and methodology, found a significant effect on plant yield as a function of the irrigation levels applied (50; 75; 100 and 150% of Class "A" tank evaporation).

According to ALMEIDA et al. (2004), the lack of plant response to increasing water levels is due to insufficient water being applied, which is evident when looking at the replacement levels used by KOETZ et al. (2006)b and SOUZA et al. (2009) and the respective results obtained. This clearly demonstrates that the levels established for the experimental intervals are important when planning research.

The regression analysis applied to the data (number of pods, fresh mass of pods and yield as a function of the water levels applied) indicated that linear and quadratic regression can be used to describe the effect of the treatments for these variables in the ranges studied (Table 3). However, the coefficients of the quadratic regression were higher than those of the linear regression, indicating that this model better represented the effect of the treatments (Table 4).

Table 4 Regression fitted to the data number (NV), fresh weight (NF) and production (DP) of pods of asparagus bean depending on the depth of water applied. **Regression fitted to the data number (NV), fresh weight (NF) and production (DP) of pods of asparagus bean depending on the depth of water applied**

Variables	**Adjusted regression**	$\mathbf{R^2}$
NV	$y = 26.6497 + 0.0295{*}Wi$	0,23
	$y = -35.6210 + 0.3938{*}Wi - 0.0005{*}Wi^2$	0,94
MF	$y = 0.00650Wi + 9.45449$	0,94
	$y = 7.74159 + 0.01653^{*} W - 0.00001 W^{*2}$	0,99
PD	$y = 246.1735 + 0.5541^{*} Wi$	0,40
	$y = - 533.02 + 5.1135{*}Wi - 0.0059{*}Wi^2$	0,96

SANTANA et al. (2009), when cultivating common beans (*Phaeseollus Vulgaris* L.) under five levels of irrigation water replacement (40, 70, 100, 130 and 160%), also noted a better fit of the crop yield data to the quadratic model.

Looking at the results shown in Table 5, it can be seen that the number of pods and yield increased until the application of 387.10 mm of water, and then decreased. The fresh mass of the pods showed a continuous growth trend as a result of the water levels applied.

Table 5 Mean values for number of pods (NV), fresh weight (NF) and production (DP) of pods of asparagus bean depending on the depth of water applied (Wi). **Mean values for number of pods (NV), fresh weight (NF) and production (DP) of pods of asparagus bean depending on the depth of water applied (Wi)**

Wi (mm)	NV	MF (g)	PD (g plant)$^{-1}$
198,8	23,94	10,60	255,18
293,9	39,44	11,30	445,45
387,1	48,65	12,33	599,46
480,7	40,67	12,74	519,30
574,4	36,53	12,93	472,25

The best results for the number of pods, production and yields were obtained with the application of 387.10 mm (100% replacement), indicating that the application of deficit (198.8 and 293.9 mm) and excessive (480.7 and 579.8 mm) levels were limiting to plant yields (Table 5). These results confirm the negative effects of inadequate irrigation water management described in scientific literature (SANTANA et al., 2009, PRAVUSHI et al., 2010 and OLIVEIRA et al., 2011).

BILIBIO et al. (2010) obtained similar results when studying the effect of water replacement rates (50, 75, 100, 125 and 150%) on the cultivation of eggplant plants (*Solanum melongena* L.), observing that the highest yield was obtained with full replacement of the water consumed.

Similarly, KOETZ et al. (2008)a, testing the effect of four levels of water replenishment (50, 75, 100 and 125%), also observed higher yields of tomato plants (*Solanum lycopersicum* L.) with water levels that maintained soil moisture close to field capacity (100% replenishment). Water replenishment above and below this value resulted in a drop in yield.

The fact that more fresh mass was obtained from the meter bean pods harvested from the plots cultivated with excessive blades (480.7 and 579.4 mm) may be due to the reduction in the number of pods produced under these growing conditions (Table 5). With fewer pods on the plants, the rate of absorption of nutrients and photoassimilates may have increased, which may have contributed to the gain in fresh mass.

Although there was an increase in the fresh mass of the pods with the increase in the water levels applied, this was not enough to compensate for the losses in the yield of the plants which produced fewer pods when grown under excessive levels (Table 5). In view of this fact, it can be inferred that the irrigation water management strategy to be adopted by the producer in order to increase the productivity of meter bean plants should prioritize maximizing the number of pods and not increasing their fresh mass.

In this study, both variable and fixed costs showed higher average costs for lower yields and lower average costs for higher yields (Table 6). This is due to the greater dilution of costs between higher yields, which reflects a more rational use of resources.

Table 6 also shows that fixed costs accounted for an average of 95.9% of total costs, while variable costs accounted for just 4.1%. As you can see, the percentage share of fixed costs is very high compared to the share of variable costs.

Table 6 Average variable cost (AVC), average fixed cost (AFC), average total cost (ATC) production asparagus bean depending on the depth of water applied (Wi). **Average variable cost (AVC), average fixed cost (AFC), average total cost (ATC) and net revenue (NR) production asparagus bean depending on the depth of water applied (Wi)**

Wi (mm)	PR (kg ha)$^{-1}$	CVMe (R$ kg)$^{-1}$	(% CTMe)	CFMe (R$ kg 1)	(% CTMe)	CTMe (R$ kg)$^{-1}$
198,8	15.310,45	0,0264	2,1	1,2036	97,9	1,2300
293,9	26.727,05	0,0223	3,1	0,6895	96,9	0,7117
387,1	35.967,74	0,0218	4,1	0,5123	95,9	0,5341
480,7	31.157,98	0,0305	5,0	0,5773	95,0	0,6078
574,4	28.335,00	0,0411	5,9	0,6516	94,1	0,6927
	Average	0,0284	4,1	0,7269	95,9	0,7553

Considering the selling price of meter beans (R$0.92 kg^{-1}) and the cost of producing 1 kg of this vegetable (R$1.23 kg^{-1}), it can be seen that when irrigation is managed by applying 198.8 mm of water during the crop cycle, the producer would make an approximate loss of R$0.32 kg^{-1} , as the selling price would not pay for the cost of production. Therefore, it would be economically unviable to produce under this irrigation management condition.

Based on the yield data, the production function was obtained by adjusting the quadratic regression (equation 15) and from these results the maximum and optimum quantities of the factors were calculated.

$$Y_e = -31.980 + 306{,}8W_i - 0{,}353W_i^2 R^2 = 0{,}96 \quad (15)$$

The quadratic term of the w_i variable was negative, indicating that the condition necessary for Y_e to have a relative maximum at the Wm point was met. The value obtained for the second derivative of the function was negative (-0.353), satisfying the sufficient condition to determine the point of maximum economic return (w_*).

Based on the adjusted regression equation (equation 15), the maximum yield (w_m) can be achieved by applying 437.46mm of water. The corresponding estimated yield will be 34,633.2 kg ha^{-1} . On the other hand, the optimum economic blade (W*) is obtained with 434.92 mm. With this value, the estimated yield was 34,629.8 kg ha^{-1} . Wm and W* are similar, a fact also observed for tomatoes (KOETZ et al., 2008), lettuce (SILVA et al., 2008; VILAS BOAS et al., 2008), beans (SANTANA et al., 2009), eggplant (BILIBIO et al., 2010) and peas (OLIVEIRA et al., 20011). This is possibly due to the high economic value of these crops, which makes it more profitable to apply water to keep soil moisture close to field capacity.

As you can see, the water saving from applying the water that maximizes the economic return (W*) is 3.1 mm ha^{-1}, and the financial advantage is 3.2 R$ ha^{-1} (Table 7), so the differences are small and irrelevant to cover the additional costs (6.3 R$ ha^{-1}) of adopting the water to obtain maximum productivity (w_m). Thus, it can be inferred that the maximum economic return can be obtained by adopting irrigation management that keeps soil moisture close to field capacity.

The small difference observed between the water quantity with which the best economic return is obtained (W*) and the maximum productivity (w_m) is due to the relationship between the cost of applying water (Cw) and the price of the product (Py), since W* depends on this relationship (equation 12). Therefore, whenever there are changes in the price of the product (Py), paid to the producer, or in the cost of the water applied (Cw), there will be a shift in the economic optimum (W*).

Table 7 Assessment of economic performance for the irrigation water that maximize productivity (Wm) and economic return (W*) for the asparagus bean. **Assessment of economic performance for the irrigation water that maximize productivity (Wm) and economic return (W *) for the asparagus bean**

	W_i (mm)	Y_e (Kg ha^{-1})	RB	C_t	RL	B/C
			--------	(R$ ha^{-1})	--------	
W*	431,3	35.156,6	31.859,4	19.300,0	12.559,4	1.7
W_m	434,4	35.160,3	31.862,5	19.306,3	12.556,3	1.7

Wi = applied blade; Ye = estimated yield; RB = gross revenue; Ct = total cost;

NR = net revenue; B/C = benefit-cost ratio.

Thus, the lower the Cw/Py ratio, the closer the W* rate is to Wm and, consequently, the lower the water savings when adopting the management to apply W* in relation to Wm. This shows that the optimum economic rate will depend on prices at the time of the economic analysis.

Table 7 shows that the economic return (R$12,559.4 ha^{-1}) for applying W* is very close (R$12,556.3 ha^{-1}) to that obtained by applying Wm, and reinforces the hypothesis that the blade with the maximum economic return (W*) is similar to the one that provides maximum productivity (Wm).

The profitability indicator (B/C) was 1.7 for W* and Wm (Table 7). This figure indicates that for every real invested, the return will be 1.7 reais. Based on the decision criteria (B/C > 1.0), this indicator shows that it is economically viable to grow meter beans using these irrigation management strategies.

4 CONCLUSION

Under the conditions in which the experiment was carried out, the results led to the conclusion that:

a) to obtain maximum productivity and economic return, irrigation should be carried out in such a way as to keep the soil close to field capacity;

b) Irrigation of the meter bean crop in a protected environment is economically viable, given the return of R$1.7 for every real invested.

REFERENCES

ARAUJO, A. P. B. de; COSTA, R. N. T.; LACERDA, C. F. de; GHEYI, H. R. Economic analysis of the recovery process of a sodic soil in the Curu-Pentecoste Irrigated Perimeter, CE. **Revista brasileira de engenharia agricola e ambiental**, Campina Grande, v. 15, n. 4, 2011.

BILIBIO, C.; CARVALHO, J. A.; MARTINS, M.; RESENDE, F. C.;

FREITAS, W. A.; GOMES, L. A. A. Production function of eggplant irrigated in a protected environment. **Irriga**, Botucatu, v. 15, n. 1, p. 10-22, 2010.

CARDOSO, M.O. **Non-conventional vegetables from the Amazon**. EMB RAPA-MA. Brasilia - DF, 1997.

CFSEMG. COMMISSION OF SOIL FERTILITY OF THE STATE OF MINAS GERAIS. **Recommendations for the use of correctives and fertilizers in Minas Gerais**. 5. ed. Lavras, 1999. 359 p.

CHEN, C.; TAO, C.; PENG, H.; DING, Y.Genetic Analysis of Salt Stress Responses in Asparagus Bean (Vigna unguiculata (L.) ssp. sesquipedalis Verdc.). **Journal of Heredity**, Hubei, v. 98, n.7, p.655-665, 2007.

DANTAS, A. A. A.; CARVALHO, L. G.; FERREIRA, E. Climate classification and trends in Lavras, MG. **Ciência e Agrotecnologia**, Lavras, v. 31, n. 6, p. 1862-1866, nov./dez. 2007.

DERMITAS, C.; AYAS, S. Deficit irrigation effects on pepper (*Capsicum annuum* L. Demre) yield in unheated greenhouse condition. **Journal of Food, Agricultural and Environment**, v. 7, n. 3-4, p. 989-1003, 2009.

DOURADO NETO, D. ; NIELSEN, D. R. ; HOPMANS, J. W. ; REICHARDT, K. ;

BACCHI, O. O. S. Software to model soil water retention curves (SWRC, version 2.00). **Scientia Agricola**, Piracicaba, v. 57, n.1, p. 191-192, 2000.

FERREIRA, D. F. SISVAR: a program for statistical analysis and teaching. **Revista Symposium,** Lavras, v.6, p.36-41, 2008.

FLYMAN, M. V.; AFOLAYAN, A. J. Effect of plant maturity on the mineral content of the leaves of *momordica balsamina* (L). and *Vigna unguiculata* subsp. *sesquipedalis* (L.). **Journal of Food Quality**, *Fort Hare, v.* 31, p. 661671, 2008.

FRIZZONE, J. A. Irrigation planning using optimization techniques. **Revista Brasileira de Agricultura Irrigada**, Fortaleza, v.1, n.1, p.24-49. 2007.

KOETZ, M.; MASCA, M. G. C. C.; CARNEIRO, L. C.; RAGAGNIN, V. A.; SENA JUNIOR, D. G.; GOMES FILHO, R.R. Industrial tomato production under drip irrigation in the southeast of Goiás. **Revista Brasileira de Agricultura Irrigada**, Fortaleza, v.2, n.1, p.09-15. 2008.

KOETZ, M.; COELHO, G.; CARVALHO, J. A.; SOUZA, R. J. de; SILVA, R. A. da. Melon production in a protected environment irrigated with different water levels. **Irriga**, Botucatu, v. 11, n. 4, p. 500-506, 2006.

MAGGI, M. F.; KLAR, A. E.; JADOSKI, C. J.; ANDRADE, A. R. S. Production of lettuce varieties under different soil water potentials. **Irriga**, Botucatu, v. 1, n. 3, p. 415-427, 2006.

MACIEL, J. L.; DANTAS NETO, J.; FERNANDES, P. D. Guava tree response to water slurry and nitrogen fertilization. **Revista brasileira de engenharia agricola e ambiental**, Campina Grande, v. 11, n. 6, 2007.

MONTEIRO, Rodrigo Otâvio C et al . Response function of the melon tree to different irrigation rates and nitrogen doses. **Horticicultura Brasileira**, Brasilia, v. 24, n. 4, 2006.

OLIVEIRA, E. C.; CARVALHO, J. A.; RESENDE, F. C.; FREITAS, W. A.

Technical and economic feasibility of pea production (*Pisum sativum* L.) grown under different irrigation rates. **Engenharia Agricola**, Jaboticabal, v. 31, n. 2, p. 324-333, 2011.

SANTANA, M. J. DE; CARVALHO, J. A.; ANDRADE, M. J. B. DE; GERVASIO, G. G.; BRAGA, J. C.; LEPRI, E. B. Technical and economic feasibility of applying water to common bean crops (*Phaseolus vulgaris* L.). **Ciência e Agrotecnologia**, Lavras, v.33, p.532-538, 2009.

SILVA, P. A. M. et al . Response function of American lettuce to water levels and nitrogen fertilization. **Ciência e agrotecnologia**, Lavras, v. 32, n. 4. 2008.

SOUSA, Antonio E. C et al . Melon productivity under irrigation and potassium fertilization. **Engenharia Agricola**, Jaboticabal, v. 30, n. 2, 2010.

VAN GENUCHTEN, M. T. A closed form equation for predicting the hydraulic conductivity of unsaturated soils. **Soil Science Society of American Journal**, Madison, v. 41, p.892-8, 1980.

VILAS BOAS, R. C.; CARVALHO, J. A.; GOMES, L. A. A.; SOUZA, A. M. G. de; RODRIGUES, R. C.; SOUZA, K. J. Technical and economic evaluation of the production of two cultivars of curly lettuce as a function of irrigation rates. **Ciência e agrotecnologia**, Lavras, v. 32, n. 2. 2008.

XU, P.; WU, X.; WANG, B.; LIU, Y.; EHLERS, J. D. A SNP and SSR Based Genetic Map of Asparagus Bean (Vigna. unguiculata ssp. sesquipedialis) and Comparison with the Broader Species. **Plos One journal**, Kingdon, v. 6, n. 1, p. 1-8. 2011.

ZHU, Y.; YU, H.; WANG, J.; FANG, W.; YUAN, J.; YANG, Z. Heavy Metal Accumulations of 24 Asparagus Bean Cultivars Grown in Soil Contaminated with Cd Alone and with Multiple Metals (Cd, Pb, and Zn). **Journal of Agricultural and Food Chemistry, [S.l.]**, v. 55, p. 1045 - 1052, 2007.

MIX
Papier aus verantwortungsvollen Quellen
Paper from responsible sources
FSC® C105338

Printed by Books on Demand GmbH, Norderstedt / Germany